NEW AGE PHYSICS

Theory of Everything Based on Consciousness, Expansion, Frequency and Density of Matter. Physics Breakthrough in UFOs, Ultraterrestrial Technology and Interdimensional Worlds

---

ROLAND MICHEL TREMBLAY

the MARGINAL

New Age Physics: Theory of Everything Based on Consciousness, Expansion, Frequency and Density of Matter. Physics Breakthrough in UFOs, Ultraterrestrial Technology and Interdimensional Worlds by Roland Michel Tremblay

Published by *The Marginal*

www.themarginal.com

Grove Cottage, Abermule, Montgomery, Powys, Wales, SY15 6NL, United Kingdom

First Edition

Version 1.0, last updated 13 December 2022

ISBN 978-1-915633-00-2 (ebook)

ISBN 978-1-915633-01-9 (paperback)

ISBN 978-1-915633-02-6 (hardcover)

ISBN 978-1-915633-03-3 (audiobook)

B&N ISBN 979-8-823117-94-4 (paperback)

B&N ISBN 979-8-823117-96-8 (hardcover)

1. UFOs and Interdimensional Beings, 2. Paranormal and Unexplained Phenomena, 3. New Age Metaphysics, 4. Alternative Physics, 5. Theoretical Physics

BISAC subject codes:

OCC025000 BODY, MIND & SPIRIT / UFOs & Extraterrestrials

OCC029000 BODY, MIND & SPIRIT / Unexplained Phenomena

PHI013000 PHILOSOPHY / Metaphysics

SCI055000 SCIENCE / Physics / General

SCI098000 SCIENCE / Space Science / General

New Age Physics book cover designed by Roland Michel Tremblay. Image/artwork used under extended licence from the creator Neven Bijelić (www.pixelparticles.com).

Section 4.9, *Figure 17 - Three different planetary orbits around two suns,* made by Roland Michel Tremblay. It is a modified reproduction of an image by Philip D. Hall, available under an Attribution-ShareAlike 4.0 International licence. His image is called *Schematic of a binary star system with one planet on an S-type orbit and one on a P-type orbit.* Own work, CC BY-SA 4.0: https://commons.wikimedia.org/w/index.php?curid=55001016

Figure 21 - Book cover of *The Final Theory* by Mark McCutcheon, photo by Roland Michel Tremblay. Published by Universal Publishers, used under fair use:

www.universal-publishers.com/book.php?method=ISBN&book=1599428660

Figure 23 - Atomic and Subatomic Expansion Equation image, made by Roland Michel Tremblay.

Photo of Roland Michel Tremblay by Patrice Bégin, used with his permission.

Destructivism: The Path to Self-Destruction and *Anna Maria* book covers, designed by Roland Michel Tremblay. Images/artwork used under extended licence from the creator John Moore (www.noir33.com).

All other figures in this book were made by Mark McCutcheon, most are from his book *The Final Theory,* and they are used with his permission (www.thefinaltheory.com). The original numbering has been kept in parentheses to help locate them in his book. They correspond respectively to the chapter and the image number in that chapter.

Contents

Foreword by Mark McCutcheon

In *New Age Physics,* Roland Michel Tremblay greatly broadens the novel notion of Expansion, driving and explaining everything in conventional science and the known universe – and beyond.

It is inspired by the two fundamental forms of Expansion in the book *The Final Theory,* coined *Expansion Theory,* covering the atomic and subatomic realms of our existence.

Atomic Expansion replaces both Newton's *attracting gravitational force,* and Einstein's *warped space-time* theories of gravity, with a singular and much simpler scientific mechanism.

Subatomic Expansion delves deeper into the subatomic realm within the atom, replacing both historic and modern models of the atom, with a consistent expansion-based model. All the while, redefining concepts such as *electric charge,* the current families of subatomic particles and anti-particles, and quantum-mechanical models.

Subatomic Expansion is further presented as the basis of other well-known phenomena outside the atom, such as electric and magnetic fields and forces, the nature and behaviour of light, and the entire electromagnetic spectrum.

As may be expected, the sweeping rethink of today's science presented by *Expansion Theory,* has the potential to lead to many scientific advances that are only aspired to or dreamt of today, such as potential solutions to the current mysteries of *dark matter* and *dark energy,* or pointing a plausible way toward faster-than-light travel and communication.

And in that spirit, *New Age Physics* casts an even broader net that captures further concepts, to extend Atomic and Subatomic Expansion from two to five forms of Expansion, presenting a comprehensive yet concise summary, relating and inter-relating the key ideas of each type.

I consider *New Age Physics* to be a prime example of my goal to inspire, such a sincere and deeply considered assimilation and integration of the original *Expansion Theory* concepts, with other theoretical physicists and philosophers out there, in order to form a unique extension of scientific and philosophical exploration. A revolution in our science awaits!

ONE

New Age Physics Introduction

1.1 *New Age Physics* - A working and credible Theory of Everything

New Age Physics by Roland Michel Tremblay, a theory of everything published in 2022, has been written for people without any background in physics, or at least with only a basic knowledge, since Atomic Expansion Theory re-writes the entire physics, and the mechanics of our physical world are now far easier to understand.

It is a credible and all-encompassing theory, uniting all four main forces of nature, through the simple principle of the expansion and contraction of matter. Two substantial books have already been written on the subject, and it has the potential to usher in a new industrial revolution. Here is an early interview on YouTube, but the book is far more up to date:

Atomic Expansion Theory Full Interview - A Theory of Everything
https://youtu.be/8lj1F8UqPc8

Essentially, we have reached the point in physics where, to keep

Einstein at our firmament, we had to invent extraordinary unlikely concepts, such as: dark matter, dark energy, gravitational waves, graviton particles, Big Bang, black holes, and singularities, otherwise none of our models would work.

The scientific thing to do, would be to declare that Einstein was wrong on all of his theories, and move on in search for new models. However, there is still nothing credible in our science, to replace him with.

We could erase Einstein and go back to Newton, but there are significant issues there too. No one has found any trace of his attracting force acting at a distance, supposed to be emanating from matter, nor its power source.

As for Quantum Mechanics, no matter how successful it is, it is largely a string of misinterpretations, of observed phenomena in physics, leading to endless claims of spooky occurrences, meaning lacking sensible explanations. It is redressed with new interpretations within Atomic Expansion Theory, from the point of view of the expansion and contraction of matter.

Einstein has put us into a black hole, for well over a century now. All the proofs out there that he was correct, have been misinterpreted by people, who only saw what they wanted to see. Humanity's future depends on a revolution in physics and chemistry, because there is a lot of potential for discoveries, that could overnight change the world for the better. There are also a lot of powerful vested interests out there, who don't want anything to change. Consequently, don't expect this to be peer-reviewed any time soon.

New Age Physics' subtitle is perhaps telling about the content: *Theory of Everything Based on Consciousness, Expansion, Frequency and Density of Matter. Physics Breakthrough in UFOs, Ultraterrestrial Technology and Interdimensional Worlds.*

At this point, what good would a theory of everything in physics be, if it still could not explain, what many people have witnessed UFOs accomplish? Or elucidate what these new age concepts, such as interdimensional worlds, are all about? How do we explain ghosts, spirits and ultraterrestrials in science nowadays? Do we?

Wouldn't it be great if we had a theory of everything in physics, that everyone could comprehend, without the need for a PhD? Which could explain absolutely everything, and most especially the four main forces of nature, which are gravity, the strong and weak nuclear forces of the atoms, and electromagnetic forces?

What if the simple new principle of the expansion and contraction of matter, could explain all these forces and more? By the end of this very introduction, you will have a complete understanding of this new physics.

However, it is not enough for a theory of everything, to unite or explain all four main forces of nature, if we cannot also unite particle physics with astrophysics. How could we use the very same equations and physics, to calculate variables, in the very small and the very large?

In a world where matter constantly expands, it is likely that all scales of reality are identical, especially when they are so similar. What we see in a microscope, must ultimately be indistinguishable, from what we see in a telescope, since the same laws of nature, drive the motion and formation of matter, no matter the size. And if we do a few adjustments to our current beliefs and models, we can see that both scales are identical after all.

Solar systems are electrons, while planets and stars are neutrinos, and galaxies are atoms. That simple idea, paired with the expansion of matter, driving the motion of everything in the universe, unite the physics at all scales, since all scales are identical.

This will be difficult to accept, because our models of the electrons and the atoms, are nothing like solar systems and galaxies. However, we have never observed them clearly, since they are too small to see.

There are irrefutable proofs that Atomic Expansion Theory is correct, and in time it will be divulged. In the book I discuss the theory itself, and the proofs that I have myself uncovered. So please, have a look at *New Age Physics,* especially the second chapter, as it contains the short and long summaries of the theory.

1.2 Five types of expansion and contraction of matter to explain the entire physics

New Age Physics is a complete theory of everything, based on five different types of expansion and contraction of matter, which should open the doors of science, to rival anything related to ultraterrestrial technology.

This includes levitating large stones and spaceships, through the buoyancy of matter of a lesser density, traveling anywhere in the universe instantaneously, and even time travel.

The same laws of physics, can now be applied to both the very small and the very large, where moons, planets and suns, are respectively sub-neutrinos, neutrinos and super-neutrinos. Anything smaller, such as asteroids, comets and rocks, are sub-sub-neutrinos. While electrons are solar systems, atoms are galaxies, and clusters of galaxies form cosmic molecules, composing objects and beings, at a yet higher scale universe.

Just like in particle physics, we will find that rogue planets orbit between solar systems, and rogue solar systems orbit between galaxies, and between clusters of galaxies, linking them together to form larger objects.

At the smaller scale, although there must be an even smaller scale universe, and despite the sub-sub-neutrinos, the neutrino is the only fundamental particle in existence, all other particles are made of neutrinos. Just like planets, moons and suns, are the only fundamental objects in the universe, since solar systems and galaxies, are composed of orbiting planets, moons and suns.

And just like planets, neutrinos are made of matter and have a mass. Any particle annihilation or destruction, does not lead to puffs of energy. It actually leads to clouds or clusters of neutrinos, and/or electrons, flying in all directions, which gather kinetic energy with motion.

As for protons, neutrons and photons, they are clusters of electrons, interlinked together through orbiting neutrinos. At the higher scale, they are clusters of solar systems, interconnected together through orbiting planets.

Within galaxies, a proton or a neutron would be a bunch of interconnected solar systems, which we might eventually be able to identify, by distinguishing individual groups of solar systems, distanced from each other.

The same is true for groups of galaxies distanced from each other, which would represent cosmic molecules. What we witness in the very small, must also be seen in the very large, and vice versa.

The changing speed at which neutrinos and electrons are orbiting, within and between larger particles, and consequently, the varying size of their orbital rings, correspond to the frequency or vibrational rate of matter, defining the interdimensional density of matter.

The five different types of expansion and contraction of matter, explain the entire physics, governing all realms of existence, even the realms of dreams and thoughts, and interdimensional worlds.

1.2a The first type of expansion and contraction of matter

The first type is the constant inner expansion of matter – the neutrinos, electrons, atoms, molecules, planets, solar systems, galaxies, clusters of galaxies, and the universe – all expand at a constant universal rate of 0.00000077 of proportional size per second square, causing the Earth to expand by 4.9 metres each second. This explains gravity and orbits, through the geometry of expansion, gravity being the changing distance, between expanding objects.

This inner expansion of matter remains unseen by us, since we also expand, along with our measuring instruments. Therefore, for all intents and purposes, there is no expansion in our resulting reality. However, although expansion only happens behind the scenes, we still see and feel the effects of this hidden expansion.

With gravity, it is the distance between the objects, that reduces as an acceleration, while the expansion of all particles and objects, remains constant. But why would the distance reduce as an acceleration, if the expansion is constant?

This is due to a further relative decrease in distance, when calculating gravity, because space does not expand along with the objects. Space always becomes relatively less, once we can no longer see the expansion. What used to be one metre of space, becomes half a metre, after the objects have doubled in size.

In our resulting reality, two objects of one metre in length each, don't suddenly measure two metres each, after they double in size. They are the same size as before, despite doubling in size behind the scenes. In order for the two objects of one metre in length each, to remain of the same length, even after doubling in size, the space separating them, must become relatively less.

This is an example of relative motion causing an acceleration, in our resulting reality, because of what is happening behind the scenes. And this is precisely why orbiting objects, such as satellites and space stations, naturally accelerate and remain in orbit around the Earth, instead of crashing down back to Earth.

As a result, while the Earth is expanding at 4.9 metres per second, the distance between the ground and a free floating object in the air, reduces as an acceleration of 9.8 metres per second square, as both the Earth and the object expand.

Objects don't all fall at the same rate. A larger falling object, will reduce the distance faster than a smaller one, because gravity is now measured from surface to surface of expanding objects, and depends on the size of the objects, instead of their mass.

Despite all the proofs to the contrary, with the limited measurements made here on Earth, even within Newton, where gravity is measured from centre to centre of the objects, and depends on mass, a more massive object should reach the ground faster, than a less massive one.

And so it is for Einstein, because his theory is equivalent to Newton through the Principle of Equivalence, and is also concerned with the mass of the objects, despite being a theory based on the geometry of spacetime, that objects simply follow.

To see this more clearly, just make the measurements of gravity using all three theories – and you can also visualise it – when the two objects falling towards the Sun, are a planet stopped in its orbit, and a marble. In all three cases, the planet will reach the Sun significantly faster than the marble.

How could it be any different, when these theories are based on the mass or the size of the objects? A planet makes a massive indentation in the sheet of spacetime, and so is the Sun. Following that steep curvature of space will be faster, than the small indentation made by the marble. Then, necessarily, mass or size must affect the result, and objects don't all fall equally, in any of our theories of gravity.

There is usually a direct proportion between the mass and the size of an object, they both depends on how much matter there is contained within the object. But it is not always the case. When it comes to calculate gravity and orbits, a planet made of Styrofoam, or made of matter of a lesser density, would bring significantly different results, between Atomic Expansion Theory based on size, and Newton and Einstein based on mass.

While using Newton or Einstein, without knowing that the planet is in fact made of Styrofoam, or is made of gases inside instead of being solid, would significantly distort the results of our calculations, because we would overestimate the mass of these planets.

With Atomic Expansion Theory, only size matters, but only as long as it is measured from orbit. Within Atomic Expansion Theory, it must be said that gravity measured on the ground, would change depending on the density of matter, and the centre of mass. And we could not tell from orbit, we would need to measure it directly on the ground.

Expansion is not uniform, it goes from the centre of mass. Because of the original formation of the Earth-Moon system, gravity will be higher on the hidden face of the Moon, because the centre of mass is nearer the Earth, and the density of the Moon becomes gradually less, the further away it is from Earth.

That the density of the Moon is not uniform, and gravity is different depending on where you land, is a prediction of Atomic Expansion Theory. Gravity on the Moon should be a quarter the one of the Earth, and overall it has to be, because it is a quarter of the size of the Earth.

Since it was measured at one sixth by the Apollo missions near side, then the centre of mass must be closer to us, the density must gradually become less, and gravity must be higher on the hidden side, up to one third the gravity of the Earth.

This might explain why several countries failed to land space probes on the Moon, if they assumed that gravity all around the Moon, was uniformly one sixth of the gravity of the Earth.

The expansion of matter also eliminates the need, for the strong nuclear force within the atom, which was required to keep the nucleus together, and prevent all the particles from flying out in all directions. Protons and neutrons in the atom's nucleus, simply push against each other, in their inner expansion.

While the weak interaction, or radiation, is due to a neutron being an unstable particle by nature, which decays into a more stable proton, an electron, and neutrinos. Particle decay is not so much a mystery, when we accept that protons and neutrons are now composed of electrons, interconnected together through orbiting neutrinos, while the electrons themselves are composed of orbiting neutrinos, and that any object or particle, can easily decompose into its building blocks, causing radiation.

1.2b The second type of growth and shrinkage of matter

The second type of expansion and contraction of matter, is the growth and shrinkage of the electrons, when the neutrinos orbiting within and between the electrons, get excited and move faster, or settle down and move slower, causing the neutrino orbits to enlarge or reduce in size.

This is the crossover effect, when electrons grow (second type of expansion) and *move out* of the atom, to push and pull against each other in their inner expansion (first type of expansion), equalising in size and numbers when in contact.

As these clouds of interconnected electrons grow in size, they can push back objects and particles as a repelling force, or they can shrink and pull back as an attracting force.

Through the geometry of expansion, and the natural orbit effect, electrons naturally move from where there are surpluses of electrons, to where there are depletions.

Electrons moving either in clouds, clusters, compressed bands, or lines on wires, goes on to explain, respectively: electromagnetism, heat and light, radio signals and microwaves, and electricity.

The first two types of expansion and contraction of matter, unite all four main forces of nature, into a theory of everything.

1.2c The third type of growth and shrinkage of matter

The third type is the growth and shrinkage of the electrons, but *within* the atom, leading to physical realities, with larger or smaller than normal atoms and molecules, more or less spaced out than usual, and of different vibrational rates and densities of matter. This is how Atomic Expansion Theory, explains other dimensional worlds and beings.

There are other realities out there, made of either larger or smaller electrons, atoms, and molecules, where neutrinos and electrons are orbiting slower or faster than normal, with smaller or larger orbital rings, which we cannot see or interact with.

But under certain circumstances, either naturally or through technology, the vibrational rate of the matter between these worlds, can equalise. Which is when we start seeing paranormal phenomena, such as ultraterrestrials, who apparently have been living here on Earth, sharing the same space as us, since time immemorial.

The process causing the expansion of all matter behind the scenes, will be discussed shortly, in the section about gravity and orbits. But for now, most of the natural acceleration within orbits, and the enlargement of orbital rings, does not affect the interdimensional density of matter, or the size of particles and objects, within our resulting reality. But a tiny part of this process is, and in time, particles and objects orbit faster, and grow slightly more, even in our resulting reality.

Consequently, as time goes by, all particles orbit slightly faster, all orbits enlarge slightly more, all particles and objects become larger, and the time rate marginally accelerates, as we constantly move into higher dimensions. Which makes any time and place in history, its very own dimension, where matter vibrates at a unique frequency, or unique resonance.

It is said that we would appear as beings of light, to people of the 18th century, because we now vibrate faster. We would also now be taller than they were, and made of matter of a lesser density.

The Earth getting slightly larger in time, albeit less dense, is responsible for the continental drift. This also explains why all continents fit together, as if they once covered the entire planet. The Earth used to be much smaller than it is today.

This is also why dinosaurs and pterodactyls, could naturally grow so large, while still being able to move and fly easily. In Atomic Expansion Theory, where gravity depends on size, a smaller planet leads to less gravity and weight, despite matter being denser. You could say that the Earth grows in time, and so are we.

1.2d The fourth type of expansion and contraction of matter

The fourth type is the instantaneous expansion, of our psychophysical individual bubble realities, from our consciousness and brain, like individual computers do on a network.

Altogether, we build an interconnected collective reality, more or less the same, and more akin to a virtual world, that we instantaneously expand out there, from our brain.

This world can only be perceived, through electromagnetic and electrical signals, received and interpreted by our brain. And we don't individually all see the same things, as can be attested by persons who are schizophrenic, or people taking psychedelic drugs.

Even hypnosis can render invisible, someone standing right in front of you. And you would be able to tell the time, from a watch held behind that person. Plus, there are several reports of mass hallucinations. As a result, any reality out there, can only exist in our mind.

And this led to the question, if there is no one in the forest, is a falling tree really falling? The falling tree only exists as a blueprint or patterns floating around. If there is no one in the woods, including animals, bugs or other living vegetation, the tree is not physically falling. But the next time we go for a walk, our brain will pick up the blueprint or patterns, and re-create the fallen tree for us to see.

It could be said that these blueprints and patterns, are like the codes of a computer programme, which we use to create reality, but only when there is a need to do so, like when we are standing there in the woods.

Our consciousness must exist outside our reality, since it is the source of our physical reality. While our brain, must be part of the illusion. And because we all have our own bubble reality, interconnected to others, each of us are the centre of our own universe.

Just like with dreams, our physical reality is more like a multitude of different bubble realities, that multiply as we make choices. Reality is much flimsier than we believe, despite seeming so permanent. But our power to change it through thoughts and will, is great, especially when done in large groups. But even individually, we are powerful.

Strong emotions, determination, and belief that we can affect reality, are key. Wishing and repeating these six things, without being specific, and without trying too hard, usually resolves everything within one's existence, after a short while: love, happiness, peace, health, wealth, freedom.

There is free will, nothing is pre-determined, and in fact, everything is constantly changing: the past, the present and the future. As the past changes, our memories of past events change with it.

This is just like in dreams, where we have previous memories, accompanying an event happening in the present, and they change depending on the event or dream. If you pay attention to your dreams, you will realise that sometimes, you do have a baggage of memories, of what happened before the moment you are experiencing. If it were not the case, it would be difficult to understand what is happening, because there would be no context.

Have you ever wondered how this was possible, and if perhaps our memories while awake, are not as real and immanent as we thought? How can we be sure that we have really lived, what we remember having lived?

This is what is meant by the expression that there is only a spacious present, or that only the present truly exists, since both the past and the future are constantly changing. This is also why there is not much point, in having regrets or remorse about anything, and why we should live more in the present moment, than the past or the future.

Any old prophecy in the scriptures, of an imminent apocalypse, could change at any time, and no longer predict any such event, through our simple belief that it will never happen.

For example, if we decide against starting World War Three, then

Nostradamus, while in the past, could never see it happening in the future. He would therefore see something else, and write his predictions differently. In the present, there would be no more prediction of World War Three for us to read about.

However, sometimes the whole idea of prophecies is to make them self-fulfilling, and were originally written, while no such events were actually in our future. A very large group of people, firmly believing that the apocalypse is coming, because they have been told that it was coming, will make it come true. At which point, the apocalypse is suddenly in our future, and the prophecy itself, has created the very event it predicted.

Prophecies are very insidious, but it also tells us the path we're on, unless we change our state of mind, which can change the past, the present and the future. We would not know that events have changed, but if we pay very close attention to it, sometimes our intuition can tell, especially when we consciously help make it happen.

This explains déjà vu, and also the Mandela Effect, which describes a situation in which a person or a group of people, have a false memory of an event, such as believing a known personality is dead, when it is not the case, while clearly remembering all the details, and when they heard the news.

The expression comes from the paranormal researcher Fiona Broome, who reported having vivid and detailed memories of news coverage, of Nelson Mandela dying in prison in the 1980s. I personally clearly remember reading all about the death of Céline Dion's husband, in a magazine in France, years before he actually died.

We need to understand how these mechanisms of existence work, because optimism, and keeping a positive outlook about life, can help a great deal to make the world a better place. At all times we are free to create the world we want, no matter the past, or any prediction or prophecy in existence.

We cannot see or interact with other dimensional worlds,

because we are not attuned to these frequencies of matter, just like a radio picks up only one station at a time. Our consciousness and brain, do not create these worlds for us to see, within our own bubble reality. Our particles would first need to vibrate at these specific frequencies, or in other words, orbit at these specific speeds.

This is why ghosts and spirits, standing in front of a crowd, can only be seen by certain people, while the others remain oblivious to their existence. And you might wonder, why a video camera could record such apparitions, if it is all in our mind.

The blueprints and patterns are there, and the consciousness of these ghosts and spirits are there as well. However, our brain will re-create it in one way, the person next to us might not re-create it at all, a mouse and a cat present at the scene will re-create it in a different way, and a video camera will record it in yet another way.

The camera is part of the re-creation of our psychophysical reality, so it can record what we create. And then, we send signals to others about what has been recorded, so the blueprint and patterns will exist, for everyone to re-create the recording in their mind, next time they watch it.

It is easier to understand how it works, when looking at it from the point of view of an open world, that people play and share on a game console, on a network. This is not to say that our reality is a computer programme, it is just very similar. But who really knows?

1.2e The fifth type of expansion and contraction of matter

And the fifth type is Universal Relativity, the expansion and contraction of matter, due to relative motion. Objects only ever expand or reduce in size as they move, in a universe without distance, where everything is located here around us.

Elephants running towards us don't cover any distance, they are simply enlarging in size towards us. Likewise, as they pass us by and run away, they are simply shrinking away from us. This would be expected, in a psychological or virtual reality, or a simulated world, or a holographic universe.

We can move anywhere instantaneously, at any particular time in history, by simply vibrating at the frequency, or resonance, of that particular time and place. In other words, by adjusting the speed and cycles of our particles, which define our relative size, our location, the time and date, and the interdimensional time rate. This is how we will instantaneously travel in the future, and possibly even travel in time.

It is also all very instinctive, often we simply need to think about it, to make it happen. We don't necessarily need to know the specific resonance of a time and place, unless we are using technology to get there. And calculating the specific resonance of a certain time and place in the past, present or future, to adjust the speed of our particles, should be within our capabilities.

This is just the core of the theory, the beginning of what it all means, and how it could change the world.

1.3 UFOs, interdimensional worlds and beings

Although in Atomic Expansion Theory, the Many-Worlds Interpretation of Quantum Mechanics is gone, there are still parallel universes, as many as our mind can create, in this psychophysical universe, shaped by our consciousness.

Our realities and planes of existence, including dreams, thoughts, and other dimensional worlds, are entirely made of neutrinos, that go on to form electrons, atoms and molecules, whether we can see these realities through our five external senses, or through our internal senses. And depending on the senses we use to look at a reality, we can perceive it as more concrete, or more intangible, when they are in fact the same.

It is said that all planets of the solar system, are inhabited by interdimensional beings. All interdimensional worlds are tangible physical realities, and are seen as such by the beings inhabiting them.

Although, looking at the fourth density world from our own perspective, would show it to be more like a gaseous reality. These beings would see the Sun, and the gas giants Jupiter and Saturn, very differently than we do. They may be more solid than gaseous, from their perspective.

We are all interconnected, on a subconscious level, through electrical and electromagnetic signals, that our brain emits and receives. Consequently, we all build relatively the same collective reality, giving it the illusion of permanence.

The real reality behind the scenes, of our world and all interdimensional worlds, must therefore be a realm of neutrinos and electrons, carrying electrical and electromagnetic signals, precisely like a Wi-Fi network does.

It is not surprising that our technology resembles the true nature of reality, since it is made of the same components and materials, and is based on the same principles and laws of nature.

UFOs can easily fly and float, and we can build anti-gravity devices, through changing the interdimensional density of matter, where objects less dense than air, are immediately pushed up above the atmosphere, through buoyancy.

Anything that is less dense, is immediately pushed up above what is denser, like a helium balloon under water, or in the air. This might even work to easily push a spaceship, out of the solar system.

A UFO can float over the Earth, like a fly can float within a moving car, with no opened windows. The air contained within, moves with the car, pushing the fly along with it, which prevents it from ending up splattered, on the back window. As a result, the fly can freely fly around, as if the car was not moving.

In its expansion, the Earth is pushing all the air upwards, pushing the UFO – of the same density as air – along with it, which is seen as anti-gravity. That same push on the air from the expanding planet, keeps the atmosphere from escaping into space, and affects weather patterns.

Therefore, UFOs don't need to use a mean of propulsion or energy to float around, once they reach the right interdimensional density. And then, they use electromagnetism to navigate, using powerful growing and shrinking clouds of electrons, to pull and push against everything else around, over long distances. When a ship weighs nothing, propelling it is effortless, and may very well be silent.

Interdimensional worlds and other planes of existence, are defined by the frequency, meaning the number of times or cycles, at which neutrinos and electrons are orbiting, within and between atoms and molecules. The interdimensional density of matter change, when the electrons, atoms and molecules' vibrational rate, size and spacing, vary, while keeping their molecular structures intact.

We know from the Tibetan Yogis, that thought and will can affect our own vibrational rate and density of matter, to the point that we can levitate in the air, and become insensitive to cold and warm temperatures.

The key to achieve it through technology, is to affect the speed of the neutrinos, so the electrons grow or shrink without leaving the atoms, and without directly affecting the atoms and the molecules.

Otherwise, what we would get if we only affect the space between the molecules, for example through heat and pressure, is simply solids moving to liquids and gases, and the molecular structure losing coherence.

Clouds of electrons, such as in electromagnetic and electric fields, capable of adding neutrinos and electrons within the atoms, or getting them out, therefore causing the growth or shrinkage of the electrons, is the answer to change the vibrational rate and interdimensional density of matter.

Dimension in this discussion, refers to the size of the particles, and not to some mathematical concept of geometrical shapes, such as hyperspheres and tesseracts. Dimensional worlds are more related to the density of matter, than seeing the world in more dimensions than height, length and width.

Although humans can travel to the fourth dimensional world, there is an issue upon the return, about remembering what occurred while there. It is just like a dream, that we immediately need to remember, or else the memory is gone. So you better document your trip, bring a video camera, and immediately review it upon your return.

Anyone experiencing time gaps, might be spending time in other dimensions, without even realising it. Not only that, but you could find yourself in the fourth density world, and although there would be a few differences, such as larger vegetation or strange animals, you might not even recognise the switch, as it might very well look like the third dimensional world.

Many people mysteriously disappear every year, and they may now be living alone in their own dimension, or in small communities of humans who also vanished, or with interdimensional beings. Thus, it could be a common occurrence, that we're just not aware of.

The Mothman, Bigfoot, mysterious beings, ghosts and spirits, different dimensional worlds, and world densities sharing the same space, are all discussed in *New Age Physics,* and are possibly now accessible to us. Is humanity truly being enslaved, by beings from other dimensional worlds? It is possible, and with this new physics, we should finally be able to fight back.

Time is not relative nor a dimension, as Einstein claimed. Time ticks the same everywhere within our own dimension, plus or minus small discrepancies, that can be explained by physical stress, and other environmental factors.

However, interdimensional time is relative, since time is relative to the vibrational rate of matter. The speed at which our particles are orbiting, dictates our time rate, or our perception of time, and the dimension we exist in.

A fly vibrating at 250 Hz, moves in fast forward, compared to us vibrating at 50-60 Hz. While tortoises vibrating at 15 Hz, move in slow motion. Flies and tortoises are moving in different dimensions than ours, ones that we can still see and interact with, and so they are still a part of our third dimensional world.

It explains why insects can walk on ceilings and walls, why flying bumblebees can support their own weight, despite their small wings, and why some spiders can fly huge distances over the ocean, without even having wings. They are slightly less dense than air, or just about the same. And the expansion of walls and ceilings, is enough gravity pushing back, for them to walk upside down without effort.

This also explains how possessed people can levitate, why demons can crawl on ceilings, and how blood or water drops can go upwards instead of downwards. That is, if we can believe everything we see in films.

Insects, animals, beings, and objects, vibrating faster or slower than flies and tortoises, can simply disappear from our view, and we can no longer interact with them. But they still share the same space as ours, just like different gases can intertwine with normal air in the atmosphere.

This explains the mysterious flying rods, only seen in photos and videos in slow motion. They are insects vibrating too fast, for us to see and interact with. And it does not take much, for even normal insects, while in flight, to simply disappear from our view. We can't even see the full motion of a bird moving its head, it seems to us to be moving in jerk moves.

It is likely that anyone who is psychic, simply has particles orbiting faster. Psychic mediums are of a different density of matter, and as a consequence, they see and interact better with the interdimensional worlds around us.

Anyone with particles orbiting faster than normal, should be able to achieve much more in a day's work, than others. Which is why we often encounter people, who can do a lot of work in a short time, which might appear impossible at first sight.

These people are not working faster than us, at least not from their point of view, they simply vibrate faster. And as a consequence, they move faster than us, and can accomplish more work than us, in a same amount of time. Just like a fly, moving four times faster, has ample time to move away, before we can catch it. Mayflies may only live a short time, but only from our point of view.

If you were to speak with interdimensional beings in the morning, and again in the afternoon, depending on the dimensional world they are in, an entire year might have passed for them. The question is, would the Earth have gone around the Sun for them, during their year?

If we are all sharing the same space, and the only difference is the speed of our particles, and as a consequence, the interdimensional density of matter, then their planets would not suddenly orbit faster than ours.

Just like for the flies, they still exist on the same Earth as us, which goes at the same speed around the Sun. If their planets were suddenly orbiting faster, their orbital rings would be far larger than ours. We could not then share the same space, and we would not encounter them so often.

We can no longer deny the existence of interdimensional worlds and beings. There is a large library of books discussing it, with countless reports, documentaries, and witness accounts. The ignorance or avoidance of the subject, and the ridiculing of the witnesses and their investigators, does not make it untrue. It is time for these phenomena to be studied more seriously, and to finally be explained by science.

1.4 Gravity, orbits, attracting and repelling forces

Gravity is the changing distance between expanding particles and/or objects. The Atomic and Subatomic Expansion Equation, can be used to calculate gravity, whether we are measuring it between two asteroids, between two sub-sub-neutrinos, or between an asteroid and a sub-sub-neutrino.

The reason is, the constant inner expansion rate, of every object and particle in the universe, is the same everywhere, no matter the scale, or the dimensional world. If it were not the case, some particles or objects, would rapidly expand out of proportion, or contract to nothingness, compared to the rest of the universe.

We might be tempted to think that, with gravity, the distance between objects always reduces, but it is not always the case. Objects could be moving away from each other at high speed, in opposite directions. And while the distance between them would still reduce, due to their inner expansion, what we would observe, is an increasing distance between them. In this case, gravity is slowing the increasing distance between the objects. Which is why gravity is a changing distance, instead of a distance decrease.

Attracting and repelling forces, are due to the changing speed and orbit's size, of smaller particles or objects, orbiting between larger ones, keeping them apart, and counteracting gravity.

For example, two solar systems are interconnected, through planets orbiting between them. As the planets and suns are constantly expanding, the distance between everything constantly reduces, which is gravity.

However, the planets are orbiting around the suns, and between the solar systems, and their orbital rings also constantly enlarge, along with the inner expansion of their components. These enlarging orbits are counteracting gravity, which is still operating in the background.

We would only notice a distance decrease between two solar systems, if they were not interconnected together, through orbiting planets between them, keeping them apart. But then, they would probably be too far apart, to make much of a difference. As soon as two solar systems get near each other, they start exchanging planets, because of the balancing act, of the geometry of expansion.

If more planets were to come into the entire system from outside, all the original planets would instantly start to accelerate, no matter where they are, because their paths would automatically follow the new geometry. Their orbital rings would also enlarge – even more than the enlarging due to the inner expansion – both solar systems would move apart despite gravity, and this would be perceived as a repelling force.

If for some reason, many planets are knocked off the entire system, the remaining planets would slow down, their orbital rings would shrink, and both solar systems would get closer together, which would be perceived as an attracting force.

You can replace the words solar systems with electrons, suns with super-neutrinos, and planets with neutrinos, and everything will remain true. If you have two clouds of interconnected electrons, all enlarging at once, or all shrinking at once, when they come into contact and exchange neutrinos, and equalise in size and numbers, then you have magnetic and/or electrical fields, that can either attract or repel other particles or objects.

This is the difference between the attracting force of gravity, and the attracting and repelling forces in magnetism and electromagnetism, which depend on a network of interconnected orbiting objects or particles, keeping larger ones apart.

1.5 The four different perspectives required to understand gravity and orbits

Both gravity and orbits are a part of the same phenomenon, and are a result of the geometry of expansion, and the natural orbit effect, where different perspectives must be considered, to understand the resulting reality:

- The first perspective is the God's viewpoint, if we could see the expansion of matter behind the scenes, where distance reduces between the objects or particles, as they expand. In this perspective, there is no gravity or relative motion affecting the objects or particles.

- The second perspective is our resulting reality, where there is no expansion that can be seen, since we are also expanding, except that we still experience the result of the first perspective: the distance reduction between objects or particles, which is gravity.
- The third perspective is the re-established expansion, but after the relative effects of the second perspective, which is useful to understand how orbital rings enlarge, and the natural accelerations and decelerations of orbiting objects or particles.
- The fourth perspective can be useful, in our thought experiments and comparisons, which is objects or particles passing each other in space, if there were no expansion, gravity or orbits.

For example, from the God's viewpoint, the expanding Moon may be moving away in a straight line, from the expanding Earth, while still going fast enough, to escape the distance reduction due to the expansion, but not fast enough to escape altogether. In effect, the distance between them remains more or less the same.

Once we can no longer see this expansion, what we see in our resulting reality, is a normal stable orbit of the Moon around the Earth, where the distance remains more or less the same.

Once you eliminate expansion, you see a natural orbit effect, where objects simply circle each other. From this perspective, it is hard to tell what might cause a natural acceleration of the Moon, in its orbit around the Earth, or to see that the orbit is actually enlarging.

In an elliptical orbit, the distance reduction, or gravity, brings back Mercury towards the Sun. But since, in the first perspective, Mercury is moving away, perhaps in a straight line, fast enough to escape the expansion of the Sun, then Mercury goes around the Sun in a slingshot effect, where relative motion accelerates it. It continues on its course, decelerating away from the Sun, until the expansion catches up with it once more, and Mercury comes back towards the Sun for another slingshot.

This is how the geometry of expansion, accelerates satellites and all orbiting objects, through a slingshot effect. And the effect is the same for circular orbits. Plus, the distance reduction between expanding objects, reduces as an acceleration, as explained previously. This shows, how orbits and gravity are linked, and part of the same phenomenon.

But from this second perspective, without expansion, the orbit seems completely stable. We see an acceleration, followed by a deceleration, that appears to cancel out the acceleration.

We don't see that, every time Mercury swings around the Sun, it gains a bit more acceleration than previously, which is the nature of the slingshot effect, and which causes the orbital ring of Mercury to enlarge.

The enlargement of the orbital rings of all the planets, causes the entire solar system to expand, at a constant rate. But remember, we don't see this expansion in the second perspective, which is our resulting reality.

To get back to the Moon in orbit around the Earth, re-instating the expansion after the relative effects, shows that the Moon, while expanding, is now orbiting in an enlarging spiralling orbit around the Earth, counteracting the Earth's expansion.

The Earth is proportionally expanding four times faster than the Moon, because it is four times the size. So to keep more or less the same distance, the Moon must be moving away in a spiral, that constantly enlarges, which corresponds to the acceleration of the orbit, which causes the expansion of all orbital rings.

From this third perspective, it is easier to understand how the orbital rings are enlarging in spirals, and that all orbits naturally accelerate or decelerate objects, depending on the geometry of the moment. There are also other phenomena, accelerations and decelerations, that can be better understood from this third perspective.

With these different and equal views of what is happening, we can explain all phenomena witnessed in the universe, not only in the cosmos – where everything moves slowly from our perspective – but also in particle physics, where everything moves extremely fast.

1.6 The best proof for Atomic Expansion Theory, the physics of a falling suspended slinky

Here is another example of seeing in different perspectives, which is in fact the best proof for Atomic Expansion Theory. You can search on YouTube for videos of a *suspended slinky falling in slow motion,* where you will see an experiment with people, holding a stretched slinky from the top of a building. When they drop it, something amazing happens, that cannot be explained by Einstein or Newton's theories of gravity.

The base of the slinky levitates in the air, immobile, until the released top reaches the base, and only then, the entire slinky starts falling to the ground. For a super large slinky, it is rather striking, it takes several long seconds before the slinky, appears to remember that gravity exists.

There are two extremely puzzling issues here. The first, when a stretched spring held at both ends is suddenly freed, both ends must come back together. One end cannot remain immobile, while the other end comes back.

The second problem is, why is the slinky levitating in the air? It should immediately start falling, even if it is stretching back, otherwise it defies gravity.

Attempts to explain this in Standard Physics range from, waiting for information to come from the top of the slinky to reach the base, before starting to fall. Or a centre of gravity nearer the bottom end, somehow makes the slinky defy gravity. Or an unstretched slinky horizontally, which does not stretch back completely, can have an end remaining immobile, while the other end is coming back, when it is hit with a stick.

This last demonstration is very different from the first experiment, where the slinky is stretched, and two forces are holding it at both ends. There is a force holding the slinky at the top, and there is a force of gravity stretching it downwards. A more accurate demonstration, would have been to pull the slinky by one end, while the rest would stretch due to the acceleration. Then, once freed, we would have seen that both ends come back together.

In all cases, the explanations are unconvincing, the slinky is simply defying gravity, by magically levitating in the air. How are we to explain this puzzling phenomenon, then?

First, since it is common sense, that both ends of the slinky are actually coming back together, then logically the bottom end must be going upwards, while the top end is going downwards. We just don't see it.

Second, since it is common sense that once dropped, gravity, or some sort of distance decrease, should instantly be happening, then logically it must be, while we just cannot see it.

Why don't we see it? Because we need to look at it from the first perspective of Atomic Expansion Theory, the God's viewpoint, where we can see the expansion.

What is really happening behind the scenes, is the expansion of the Earth is pushing upwards the person holding the top of the slinky, and the stretch corresponds precisely to this expansion rate. Therefore, when the slinky is dropped, the bottom end is moving upwards, at the exact same rate as the expansion of the planet.

This is why the base of the slinky appears to be levitating in the air immobile, when in fact, the bottom end is going up, while there is actually a distance decrease immediately happening.

The expanding ground is approaching the slinky, while it is free floating in the air, so the slinky does not defy gravity. However, it can only be seen and understood from the first perspective, where we can see the expansion of the planet.

This experiment clearly shows how a theory of gravity that is based on geometry, exerts no force whatsoever on any falling object. They simply float there in the air, while expansion catches up with them. Raindrops and snowflakes don't fall to the ground, the Earth expands towards them.

Einstein's Theory of General Relativity is also geometry based, and we could be tempted to think that, in this experiment, the slinky is likewise free floating in the air, while the base is going up, and therefore, this is the reason why the slinky is not falling immediately. But this would still be a magical levitation defying gravity.

With Einstein, the raindrops and the snowflakes immediately fall to the ground, following the curvature of spacetime. And Einstein's Principle of Equivalence to Newton, means that it is equivalent to a force acting upon them. The curvature of spacetime is like a steep slope, that objects instantly follow. Even if the base was going up, the entire slinky would immediately start falling.

Remember, the slinky is perfectly immobile in the air for several seconds. Only the expansion of the Earth behind the scenes, at the same rate that the base is moving upwards, can perfectly explain what we see.

And more importantly, we can see that gravity works immediately. There is an instant distance reduction between the slinky and the ground, and we simply don't see it, because the slinky is moving upwards at the same rate.

Gravity and orbits don't wait for information, that takes time to arrive from elsewhere, before they start happening when the geometry changes. Like it would be with Einstein, where the slinky would refuse to go down a steep slope, for several long seconds, while the slinky is stretching back. There is no clear explanation within Einstein, to explain why the slinky would defy gravity, and magically levitate in the air.

1.7 The formation of planets, solar systems, galaxies, neutrinos, electrons, and atoms

And now, for the last example, let's have a look at how planets, solar systems, galaxies, neutrinos, electrons and atoms form.

In all cases, it is just a question of a lot of matter agglomerating in a same location, through the expansion of everything, and all distances decreasing, until a massive ball of matter collapses, under its own weight and pressure. It then erupts, through internal matter squeezing out at both poles, following the path of least resistance.

These two jets of matter, are the reason why solar systems, galaxies, electrons, atoms, and the rings of Saturn, are mostly flat objects. Although, when they move extremely fast in the very small, they may appear to be spherical, while they are not.

And of course, there are exceptions, and some galaxies are not flat. They must have had jets of matter ejected in more than two directions, or more likely, they collapsed with other galaxies.

And then slowly, with the expansion causing motion, natural orbits form, and all the matter shapes up into orbiting components. For solar systems and electrons, which move more rapidly, the orbiting expanding matter shapes up into orbiting planets and neutrinos. But for galaxies and atoms rotating more slowly, the two jets of ejected matter at both poles, form into long arms. And a further explosion, might add two more arms in time.

The closer to the galactic centre, the faster matter orbits, just like planets closer to the Sun, orbit faster than the ones further away. Consequently, in time, the galactic arms twists into spirals, because matter nearer the centre orbits faster, while matter further away orbits slower. Eventually, because of the expansion, the matter of the arms form into solar systems.

Matter is not moving towards the galactic centre in a galaxy, like water in a sink hole. On the contrary, the orbits of the solar systems in a galaxy, are very much like the ones of the planets in a solar system. They are all in stable orbits, although moving together in arms, because many are interconnected through an exchange of orbiting planets. And behind the scenes, in the third perspective, their orbital rings are constantly enlarging in more spirals.

There are no more black holes, singularities or worm holes in Atomic Expansion Theory. The centre of a galaxy is either a massive bright sun, or a massive dead sun, which simply does not emit any light. Either way, it is just a big expanding lump of matter, kept at a distance from the orbiting solar systems, by their enlarging orbital rings.

If there are any stargates out there, these devices don't bend space by opening worm holes. They must either accelerate or decelerate the speed of your particles, to relatively expand or contract you, to the location you want to be. Distance does not exist, everywhere is around you, either very large or very shrunk.

1.8 What is truly relative in physics, and the end of the space-time-mass-energy concept

The elimination of Einstein's theories from our science, leads to a breakthrough in faster than light travel and communication, and the search for extraterrestrial intelligence (SETI).

The reinterpretation of quantum entanglement, according to Atomic Expansion Theory, suggests that particles from a same source, remain connected through a bridge of electrons, over long distances.

Like for a Newton cradle, knocking the first electron or particle, instantaneously affects the entire line of electrons between the particles, providing faster than light communication along any beam of light, such as the ones from the Sun, the Moon and other stars. Some extraterrestrial species from the third dimensional world, might already be using such a mean of communication, which we might now be able to access.

There is no more speed of light limit. Provided we can find a mean of propulsion, capable of accelerating us gradually for a very long time, then there is no reason we can't go at multiple times the speed of light. And if the acceleration and deceleration are slow and gradual, we won't have to worry about crushing G forces either.

This could be tested by designing a miniature ship with its own ion engine, or capable of ejecting clusters or clouds of electrons as propellent, while absorbing that energy from surrounding fields. If we then placed the ship in a particle accelerator, the push from the electromagnetic fields within the tube, could get the ship to reach a very high speed. And thereafter, its own engine could push it beyond the speed of light.

Within a particle accelerator, the ship could never exceed the speed of light, without its own mean of propulsion, because the electromagnetic fields, cannot push the ship any faster than their own maximum speed. But the fields can provide an inexhaustible mean of propulsion, and keep the ship away from the walls, so it can be accelerated to faster than the speed of light.

Contrary to what Einstein postulated, neither the ship nor the particles would become more massive with acceleration, to the point of requiring an infinite amount of energy, for the ship to overcome the light speed barrier.

One day, we might build a massive electromagnetic particle and object accelerator in space, capable of accelerating and launching spaceships and probes, at faster than the speed of light. The ships and probes would only need to use their mean of propulsion to decelerate, well before they reach other solar systems.

If we build similar accelerators in these other solar systems, travel in space within our own dimension, could become practical for colonisation. This method would also be useful within our own solar system, for travel at sub-light speed. It might prove more efficient and direct, than acceleration through slingshot effects, around planets and moons.

How we will see an event from the light coming towards us while travelling in space, will be either in slow motion if going lower than the speed of light, frozen in time if going at the speed of light, or in fast forward if going faster than light.

It depends on our speed and direction of motion, and the ones of the source of the light. None of it affects the event being observed, or the time rate on the ship, or the time rate where the event occurs. Time is now constant, the time rate is the same everywhere, within our own dimension.

We are likely to see multiple occurrences of a same event, when going faster than light, while moving towards the source of the light. Because we catch up rapidly with countless clusters of electrons, all showing an event as it occurs in time. And if the object observed is moving rapidly, then we will encounter all at once, several clusters showing us the object, in different locations.

We might even see this well before we reach the speed of light. Because not all clusters of electrons move at the same speed, since the speed of light is no longer constant. The speed of light is now variable, and subject to relative motion. That might be why it is thought, in Quantum Mechanics, that a particle can be at more than one location at the same time, when in fact, it is never the case.

The speed of light is now defined, as the changing distance between the source of the light, and the destination we're measuring it from. Meaning, it is a calculation of gravity, using the Atomic and Subatomic Expansion Equation.

The speed of light is the time that it takes, for the expanding clusters of electrons, to reach their expanding destination, after the initial speed boost required, for the clusters to leave the atomic world, which accounts for most of the speed of the light.

There are other factors that accelerate and decelerate light, and makes its path deviate, on top of the simple dissipation of the clusters of electrons, as they encounter other particles along the way. It is no simple calculation, but it is far from constant.

And it does not matter where we are, or at which speed we are going at – even faster than light – it is still just a measurement of gravity, between the source of the clusters and the destination, after having established the initial speed of the clusters, when they were thrown out of the atoms. Hence, even while going faster than light, light will still arrive at us at a specific speed, and we will still be able to see just fine.

With relative motion, our visual variables, such as light and the size of objects, become relative and changing, when compared from one frame of reference to another. But it is only an optical illusion, due to the time that it takes, for the clusters of electrons within light, to reach different locations.

The normal size of the objects doesn't actually change, and we don't age faster, or truly freeze in time with acceleration, or near a black hole because of gravity, as Einstein claimed, no matter if we go faster than light.

$E=mc^2$ is just the equation for the kinetic energy of a moving particle or object, such as the clusters of electrons within light, known prior to Einstein. There is no more space-time-mass-energy concept, all interlinked together and affecting each other. Time is no longer relative, it runs everywhere at the same rate. And space is distinct from time, one does not affect the other.

Mass is the measure of the amount of matter within a body, a particle or space, usually measured in kilograms. But in the context of Einstein's relativity and $E=mc^2$, we are talking about relativistic mass, which increases with velocity. This concept of relativistic mass and energy is incorrect.

What really increases, is not the mass or relativistic mass of an object, it is the kinetic energy of the object. Matter does not gain more mass with motion, but it does gain kinetic energy. The higher the momentum, the higher the impact on objects.

The changing speed of matter, should not affect its mass, although it affects gravity and orbits, attracting and repelling forces, and other energy measurements. But mass does not become energy, and energy does not become mass.

Mass and energy are not relative either. These values don't change, just because different persons, are observing the event from different locations. These values are not visual ones, that take time to travel to all observers, at the speed of light. So they are not relative values.

The speed of light is no longer constant. The speed of light is relative and variable, even in a vacuum, while being subject to gravity and acceleration. Light is made of clusters of electrons, it is made of matter and has mass.

As such, there is no difference between a light beam, and bullets being fired from a gun. There is an initial high-speed boost, when the gun is fired, or when a light is turned on, and thereafter, gravity affects the electron clusters, just like it would affect the bullets, or hurtling asteroids in space.

The speed of light is relative, because any moving matter, is subject to relative motion. With relative motion, you get natural accelerations and decelerations, and path deviations, such as the slingshot effect, as objects move across expanding larger objects in space. This is due to the geometry of expansion, and the natural orbit effect, and not to a twisted sheet of spacetime, that objects follow.

So, not only the speed of light is not constant, but it can also seriously accelerate, through the slingshot effect, which is light entering into a partial orbit, around a very large object, but moving too fast, to actually enter into orbit. It creates a significant path deviation and an acceleration, as can be seen in the gravitational lensing effect.

It was always clear in our science, that light bends due to gravity, and that it slows down and dissipates, as it crosses gases, liquids, and transparent solids, such as glass. Was it not logical then, that its speed was also affected by gravity, instead of just its kinetic energy measurement?

Something must be hitting something, for kinetic energy to exist in the first place, so there must be matter and mass involved in light. It is impossible for matter or mass not to be affected by gravity. And since gravity is an acceleration, then the speed of light must accelerate, even in a vacuum.

The only reason it seems counterintuitive, is because for well over 100 years now, we have been told that the speed of light is constant, with countless so-called proofs confirming this.

We've also been told, that light simply follows a spacetime medium that is warped and curved, all without affecting its speed, because it has no mass. How it can acquire kinetic energy then, is a mystery. All the proofs that the speed of light is constant, are simply due to the limitations of our experiments.

We ignored all the evidence proving otherwise, by repeating the mantra, that light is only constant in a vacuum. Here is some news, although space is said to be a vacuum, it is not devoid of matter. Especially when light are clusters of electrons, and electromagnetic fields are clouds of electrons, not forgetting countless neutrinos, permeating all of space. Do real vacuums even exist?

We missed this in Standard Theory, because the concept of light as photons, within Einstein, meant that light was an ethereal concept, with no mass involved, as if it never truly existed in the first place. Just like electrons really, supposed to be point particles, constantly appearing and disappearing, which don't have a volume, a mass, a weight, a shape, or a tangible existence.

These arguments must have only come about, because of the limits of what we can see and detect, with our instruments. We must have got lost, in our equations and theories, abandoning common sense. And we could not imagine, that electrons are just like solar systems, and that photons could be made of clusters of electrons, which have a mass that we are still unable to detect.

The big clue was there all along, anything that can have an impact when it hits something, must be made of matter, and therefore must have a mass. And anything made of matter, is affected by gravity and relative motion.

Energy within Atomic Expansion Theory, is always matter shaping up into particular forms, such as particles, clouds, clusters, compressed bands, or lines on wires, which move according to certain principles and laws.

Nothing is intangible or ethereal, or non-existent, while still having an impact on objects and particles, such as in kinetic energy. Energy is always expanding matter in motion, even when in clouds of electrons, in a static electrical field.

And the conundrum of particles moving sometimes as particles, and at other times as waves, is easily answered. Sometimes matter moves as a sea of particles moving in waves, and at other times, as a sea of particles moving, but not in waves, just like the behaviour of droplets of water composing an ocean.

To conclude, it is not time that is relative, time is constant. It is the speed of light that is relative, because it takes time for light to get somewhere, and motion is relative to the point of view.

Relative motion makes moving objects, appear larger as they come towards us, or smaller as they move away, but once again, this is all just an optical illusion. Size does not change with motion, it only changes relative to the viewpoint of different observers.

When you think about it, the fact that there is no real distance being covered in the universe, and that everything is located here around us, is a natural and logical conclusion, of motion being relative to the point of view.

And it all makes sense, if reality is simply a psychological construction, built by our consciousness, and interpreted by our brain, not unlike what computers do. There is ample evidence that reality is psychological in nature, while still being made of matter, hence the expression psychophysical reality. And finally, in a psychological reality, or a virtual one, there is no real distance, there is only the illusion of it.

Even if this reality was to be as real and permanent as they come, and you were to believe that it was not psychological in nature, then only the fourth type of expansion and contraction of matter would go, and everything else stated here, would still stand.

Don't write off everything, just because certain things seem unconvincing, at this time of our understanding. Let's hope that one day, an army of people will work on developing these ideas further, so humanity can finally reach its full potential.

1.9 Debunk of Standard Theory

Redshifts, the double slit experiment, quantum entanglement, the Heisenberg uncertainty principle, and other proofs of Einstein's theories, have all been misinterpreted. This is not surprising, when these theories and hypotheses, were initially built upon false premises.

Although the following free excerpts from *The Final Theory* by Mark McCutcheon, don't discuss Atomic Expansion Theory, they help identify where our science has gone wrong, and what a theory of everything must answer:

Debunk of Einstein's Special and General Relativity, and Quantum Mechanics
www.themarginal.com/Relativity_QM_Debunk.pdf

Debunk of Newton's Gravity and more about General Relativity (Chapter 1 of *The Final Theory)*
www.themarginal.com/FinalTheoryChapter1.pdf

Debunk of Dark Matter, Dark Energy and the Big Bang (Excerpt from *The Final Theory* titled *Cosmology in Crisis)*
www.themarginal.com/CosmologyInCrisis.pdf

Pioneer Anomaly, Slingshot Effect and Gravitational Inconsistencies Explained
www.themarginal.com/pioneer_anomaly.pdf

If you read both *New Age Physics* by Roland Michel Tremblay, and *The Final Theory* by Mark McCutcheon, you will never see the world the same way again.

TWO

New Age Physics - Short and Long Summaries

The introduction, and the short and long summaries, all review what is in the book. There are repetitions, although each section contains unique material. These are concepts not easily explained, or grasped, and every time there are more details, and it is stated differently, as to help the understanding of these concepts.

2.1 *New Age Physics* Short Summary

New Age Physics is a complete theory of everything based on five different types of expansion and contraction of matter, which should open the doors of science to rival anything related to ultraterrestrials technology, including levitating large stones and spaceships through the buoyancy of matter of a lesser density, traveling anywhere in the universe instantaneously, and time travel.

The first type of expansion is based on Mark McCutcheon's Atomic Expansion Theory in his book *The Final Theory.* It presents to the world a new theory of everything that works out of the box, to replace Newton's Law of Universal Gravitation, Einstein's General

Theory of Relativity and Quantum Mechanics. The theory is based on the one principle that matter simply expands at a constant rate, reducing the distance between everything, explaining gravity.

Falling objects are not falling, they are free floating in space until the Earth catches up with them in its expansion. We stay on the ground because Earth is expanding underneath our feet, pushing us upward. This expansion is unseen, because we are proportionally expanding as well, along with our measuring instruments.

Protons and neutrons, just like all particles, are now composed solely of expanding electrons. In their own inner expansion, the protons and neutrons are pushing against each other within the atom, eliminating the need for the strong nuclear force to prevent them from flying apart. A neutron is an unstable particle by nature that decays into a more stable proton, releasing an electron and neutrinos in the process, and this explains the weak nuclear force, or radiation.

For Mark McCutcheon, the electrons are bouncing off between the expanding nuclei of atoms, justifying chemical bonds, although for me, the electrons are still orbiting. Planet orbits are described by an ingenious new natural orbit effect, through the relative motion of the geometry of expansion.

Light and other radiant forms of energy are composed of different sizes of clusters of electrons. Electricity is a flow of expanding electrons around a wire, pushing against each other in their expansion. While electromagnetic fields are clouds of expanding electrons, pushing or pulling each other, adjusting in size and numbers. Radio waves and microwaves are compressed bands of electrons freely expanding into space.

This is all four main forces of nature explained, while they are no longer forces per se, now that we understand the true nature of expansion. All of this is covered in more details in the *New Age Physics Long Summary*.

The very small is identical to the very large and obeys the very same laws of physics. *New Age Physics* presents a new model for the sub-neutrinos being moons, neutrinos being planets, super-neutrinos being suns, while electrons are solar systems, and atoms are galaxies. The electron now has its own nucleus with orbiting neutrinos, with neutrinos also orbiting between the electrons, bonding them together to form particles, clusters and clouds.

Solar systems are linked together by planets orbiting around and also between solar systems, far from the nucleus, the nucleus being our eight planets and the Sun. There is no reason for these orbiting planets to come anywhere near our eight known planets. Galaxies are linked together by solar systems orbiting between them far from the nucleus, the nucleus being the entire visible galaxy. And finally, bonded galaxies form larger cosmic molecules, which are also bonded together by orbiting solar systems.

The second type of expansion, the crossover effect, concerns the growth and shrinkage of the electrons based on the speed at which the neutrinos are orbiting within and between the electrons. Exciting the neutrinos affects their speed, and enlarges or contracts the orbits of the neutrinos, causing the electrons variable growth or shrinkage, and causing the electrons to escape from the atoms in the crossover effect.

This interrelation of neutrinos between electrons is what is referred to as their charge, positive or negative, as the number of neutrinos contained within electrons and between them, will balance out or push back, causing an attracting or repelling force between them and other particles.

There are no more charges or charged particles in electromagnetism, just like there is no more force acting at a distance or mass involved in gravity, and yet the entire energy spectrum can be explained through the crossover effect. A simple process of growth and shrinkage of electrons pushing against each other, constantly equalising in size, which also causes their motion. They move from where there is a surplus of electrons towards where there is a depletion, in a constant search for balance in size and numbers.

The third type of expansion also concerns the growth and

shrinkage of the electrons, but within the atoms instead, without the crossover effect ejecting the electrons outside the atomic world. When neutrinos are excited, their orbits around and between the electrons enlarge, causing the growth of the electrons orbiting within and between the atoms. In turn, this results in the atoms and the molecules vibrating faster, growing in size, and their spacing enlarging, along with the orbits of the neutrinos and of the electrons. Essentially, this creates atoms and molecules of different sizes, more or less spaced out, creating different densities of matter, and defining other dimensions and planes of existence. This leads to a second type of density, which I have called the interdimensional density of matter.

This finally explains the New Age concept of matter and energy vibrating at different rates and frequencies, and worlds of other dimensions and densities. There are physical worlds out there made of smaller and larger atoms and molecules, differently spaced out, that we cannot see or interact with, except in some unusual circumstances. These worlds are intertwined with ours like the gases are surrounding us. Except that their particles are not volatile, they keep their own internal molecular structures.

The usual type of density of matter, which only concerns the spacing between the molecules as they become more excited, defining solids, liquids, gases and plasmas, also exist in other dimensions. The difference is that the atoms and the molecules don't change size or their vibrational rate, within their own dimension, and neither for the electrons while within the atoms. Outside the atoms, within the second type of growth and shrinkage of matter, electrons will still grow and shrink within their own dimension. For people with particles orbiting slower or faster, their world looks just as solid and physical as our world looks to us.

It is said that most alien species live in the fourth density, they have always been here, for much longer than we have. They even have cities everywhere, that we cannot see or interact with, because they exist and move in fast forward. The episode *Wink of an Eye,* of the original Star Trek series, gives us a good idea of how it works. It shows the Scalosians, moving too fast to be seen or heard, other than a faint buzzing sound, boarding the USS Enterprise and abducting Captain James T. Kirk.

The fourth type of expansion concerns Jane Roberts' book series *The Early Sessions,* where our reality is psychological in nature, or psychophysical, and is instantly expanded out there from the centre of our brain every moment, very much like a simulated virtual reality created by a computer. We each live in our own bubble reality, and we all sensibly build the same shared reality through telepathic connections, very similar to an interconnected network of computers.

The fifth type of expansion, is the Universal Relativity Theory, replacing Einstein's Special Relativity. It is a theory of relative motion based on the expansion and contraction of matter, instead of movement in any real space. Reality being psychophysical in nature, it is like a virtual, simulated or holographic universe, where everything is located here, and everything in the distance is simply shrunk from our point of view.

As something is moving towards us, it is only expanding towards us, just like it would be in a computer simulated world which would not occupy any real space. And when that object is moving away, it is only shrinking from our point of view, it is not covering any distance. And this is how we can instantly travel anywhere in the universe, by shrinking ourselves to that location, through changing the orbiting speed and size of our particles, or our vibrational rate or resonance.

There is no more speed of light limit, as this limit only came from Einstein's concepts of spacetime and relativity. It is not time that is relative, it is the speed of light, because only matter can be affected by relative motion. Any spaceship going at multiple times our actual measurement of the speed of light, will still see just fine. Although they might see in fast forward or slow motion, and even freeze what they are looking at, depending on their direction of motion and their speed. Their measurement of the speed of light might be similar to ours, but compared to our measurement, it would actually be multiple times the speed of light. This is what is meant by the speed of light is relative to the frame of reference.

The speed of light, and any electromagnetic phenomenon such as radio signals, is not constant. Most of the speed of light follows from its original ejection from the atoms, and from the subatomic world, where the clusters of electrons must be excited enough to reach a certain speed, in order to detach themselves and move freely into space. It is like bullets being fired from a gun, the clusters of electrons get an initial high speed boost, followed by any acceleration due to gravity.

Light can be slowed down as it crosses gases and magnetic fields, but it can also be accelerated further through the slingshot effect, whenever its path deviates through partial orbits around large objects, such as the Sun and the planets. The speed of light is variable, whether it is in a vacuum or not, and despite all the proofs to the contrary. Because light is made of matter, clusters of electrons, and all matter is subject to gravity, the geometry of expansion, including partial orbits, and relative motion.

Even within the concept of light being just photons, or packets of energy, light still enters into partial orbits around large objects and deviates from its path, as can be seen in the gravitational lensing effect. This is hardly a parallax effect or an optical illusion, the path of the light physically changes, precisely like the path of a bunch of asteroids would. This demonstrates that light is subject to gravity and to the slingshot effect, and we cannot escape the fact, that both of these phenomena create an acceleration. Consequently, we better think of better ways to test the speed of light and radio signals.

Although the Earth and the Sun are constantly expanding, reducing the distance between them, they follow the natural orbit effect of the relative motion of the geometry of expansion. Their distance keeps reducing and enlarging constantly between them, because they are going fast enough to escape their respective inner expansion, moment by moment.

Behind the scenes, if we could see the expansion, the Earth might be moving away from the Sun in a straight line, where there is no orbit at all. In the resulting reality however, where we cannot see the expansion, if the Earth is not going fast enough, it will crash into the Sun. If the Earth is going fast enough, it will enter into a stable orbit, following the geometry of expansion. And if the Earth is going very fast, it could escape the solar system altogether. The speed of the planets defines how far away they will orbit the Sun.

All the while, the Sun, the planets, and their overall orbital rings around the Sun, are also expanding at a constant rate. It explains why the reduction in distance between everything, does not affect the relative distance between all their orbital rings. The planets true orbits that we cannot see, from this point on, are in fact enlarging spirals around the Sun, counteracting their mutual expansion.

The relative motion of the geometry of expansion accelerates orbiting objects, which causes their orbiting spirals to enlarge. We cannot see any distance shrinking in our solar system, because everything remains more or less at the same relative distance, in the overall proportional expansion of everything. However, we can witness the effects. For example, two expanding solar systems would constantly come closer together, because of this overall distance reduction, due to the inner expansion of everything. This said, even these expanding solar systems are being kept apart by planets orbiting between them.

Orbits are calculated using Kepler's laws of planetary motion and a newly rediscovered Geometric Orbit Equation, which was embedded in Newton's Orbit Equation. Gravity depends on size measured from surface to surface of the objects, not on mass measured from centre to centre. And orbits are entirely a feature of the relative motion of the geometry of expansion, no mass or force acting at a distance are involved.

There are more planets as yet undiscovered orbiting very far from our eight planets, probably reaching distances equal to in-between solar systems. These planets can easily orbit two solar systems, linking them together if there is an imbalance of planets, which would translate into an attracting force. Or the solar systems can repel each other if they have the same number of planets, which would translate into a repelling force.

The speed and size of these planets' orbits between solar systems, can also bring them closer together or further away from each other, depending on how much matter is being added or is moving away from the entire system. Which can also be described as an attracting or repelling force, usually measured by Coulomb's Law in the very small.

In static electricity, Coulomb's Law is like measuring the orbits of neutrinos within and between growing and shrinking electrons, and the orbits of electrons within and between atoms. If one electron has fewer orbiting neutrinos, or one atom has fewer orbiting electrons, they will attract other particles with more neutrinos or electrons. If they have the same number of orbiting particles, they will repel. And the force of attraction or repulsion is what Coulomb's Law is meant to measure.

If we could measure such orbits, we would be using Kepler's laws of planetary motion in the very small as well, but that might prove impossible at that scale and speed. But possibly Coulomb's Law could be used or adapted to the larger scale, now that we have identified that solar systems are electrons linked together by rogue planets, comparable to neutrinos.

In gravity there is a distance reduction between objects (so-called Newton's attracting force) due to the inner expansion of all particles and objects, whether they are made of atoms or of electrons. And there is a different attracting and repelling force (similar to Coulomb's Law) due to the changing speed and size of the orbits of smaller particles or objects, linking larger ones together.

This is what is referred to as a positive or a negative charge.

Although, sometimes, a simple natural orbit effect between two particles, or the failure of one materialising, can also be interpreted as charges, or attracting and repelling forces between particles. There could also be an exchange of several neutrinos, even when dealing with only two particles such as an electron and a proton.

Examples of smaller objects linking together larger ones, are neutrinos bonding electrons together, electrons bonding atoms, electrons bonding molecules, while the atoms are composing the molecules. At the larger scale, planets bonding solar systems together, solar systems bonding galaxies, solar systems bonding cosmic molecules, while galaxies are composing the cosmic molecules. Clouds, or even just a few of these smaller particles, can either push back or pull together the larger objects, or groups of them, through the growth and shrinkage of their orbits. Again, this is how attracting and repelling forces are explained in Atomic Expansion Theory.

Time is no longer linked to space as in spacetime concepts, nor is time relative or a dimension of its own as Einstein described it, as there is no real space nor distance. There is a new concept of the relativity of time, based on the rate or frequency at which electrons orbit within and between atoms. It is the concept of interdimensional time, where one can accomplish more within a defined amount of time, if he or she is vibrating faster.

Very much like flies vibrating at 250 hertz have the time to move away, before a human vibrating at 50-60 hertz can catch them, as humans move in slow motion from the flies' point of view. Just like tortoises vibrating at 15 hertz are moving in slow motion from our viewpoint. Flies are living in a higher dimension than ours, one that we can still see and interact with. Their particles are moving faster than ours, they are larger and more spaced out, and they are less dense than we are.

Things that are less dense can levitate easily, because what is less dense is automatically pushed up by what is denser, like a balloon under water. This is why insects easily walk on ceilings, why bumblebees can defy the laws of physics and support their own weight, despite very small wings, and why spiders without any wings can fly great distances over the ocean. With a flying saucer made of a matter less dense than the air, you can easily levitate off the atmosphere without any need for propulsion.

To travel somewhere, it is like tuning a radio or a TV to the right interdimensional frequency of matter. Every place and time in history has its own vibrational rate or resonance. To go to any place at any time in history, you only need to vibrate at the right frequency or resonance of that time and place. It is all a question of expansion and contraction of matter, the size of the particles, which depends on the speed or frequency of all the orbiting particles.

If aliens and ultraterrestrials can go at any time in history, in a past and a future that are both always in movement and changing, affecting the present, then we can understand why they don't see the importance of time running as a linear thing like we do. They go through reality more like in a dream world, or something imagined and created out of our minds, where time and space are meaningless. And now we have the new physics required to explain all of this.

And to finish, there never were any dark energy, dark matter, graviton particles, gravitational waves, Big Bang or singularities to be found. They were only required for Einstein's equations to still work and match observation, when we should have admitted long ago that he was simply incorrect.

And there are no black holes either. Black holes are simply massive dead stars which don't emit light, therefore we cannot see them, while we can still measure their effects on gravity in their surroundings. We can forget about wormholes in order to travel somewhere, however, now there is no more distance. The entire universe is here with us, just shrunk from our point of view.

The entire physics can be explained by five types of expansion and contraction of matter, and how many and how fast neutrinos and electrons, and planets and solar systems, are orbiting everywhere. And now I invite you to check out the long summary, which is the rest of chapter two.

2.2 *New Age Physics* Long Summary (the remainder of chapter 2)

New Age Physics is a solid new science based on Mark McCutcheon's Atomic Expansion Theory in his book *The Final Theory.* Along with my own concepts and my Universal Relativity Theory, everything in physics can now be explained by five different types of expansion and contraction of matter. Altogether this is a full theory of everything which should open the doors of science to rival anything related to ultraterrestrials technology, including levitating large stones and spaceships, traveling anywhere instantaneously, including to higher dimensions, and time travel.

2.3 The constant expansion of all atoms is gravity

The first type of expansion of matter is the one of Mark McCutcheon, in his theory of everything detailed in his book *The Final Theory,* which re-writes the entire physics based on the single principle that atoms are expanding at a constant rate. The expansion of the atom is gravity.

Gravity is the changing distance between objects, due to the inner expansion of matter. There is an absolute decrease in distance, because of the inner expansion of the objects themselves, and an added relative decrease in distance causing the acceleration, due to space not expanding along with the objects. This is all explained visually in detail in chapter 3 of *New Age Physics,* in section *3.5 Two types of distance decrease to calculate gravity.*

This expansion is only happening behind the scenes, it is unseen by us, because everything is proportionally and relatively expanding at the same constant rate, us included. In the resulting reality from what is happening behind the scenes, there is no expansion per se that we can see. For example, nothing is expanding around us. And yet, we can witness all the effects of this hidden expansion, such as gravity.

We stay on Earth instead of floating into space, because the Earth is expanding underneath our feet, pushing us upward and keeping us on the ground. The constant expansion of the planet is also pushing the atmosphere upward, preventing it from escaping into space, while affecting the weather patterns. Rain and snow are not falling, the Earth catches up with them in its expansion. This is a complete novel way of looking at the world around us.

An object falling from the sky is free floating in the air until the Earth, in its expansion of 4.9 metres per second, catches up with the object. The relative decrease in distance makes it an acceleration at 9.8 metres per second square, as confirmed through observation. Using these very numbers, Mark McCutcheon derived that the universal atomic expansion rate, and of everything else in the universe, is a constant 0.00000077 of proportional size per second square.

Behind the scenes, all the planets' orbits around the Sun, and the moons' orbits around the planets, are enlarging in spirals, compensating for their mutual expansion. The amount of enlargement of the orbits depends on the speed and the size of all the astronomical bodies within the system, and on the geometry of expansion.

As all objects expand at a constant rate, and all orbits cause an acceleration of all objects, as I will explain later, there will always be at the base a certain amount of enlarging of all orbits. This leads to an overall constant expansion of the solar system.

And then, on top of this constant expansion, the solar system and all its orbits can grow further, or shrink back, depending on other astronomical bodies entering or leaving the solar system. More matter being added to the system affects and speeds up all the orbits, causing the solar system to grow further. The orbits could also be further affected by astronomical bodies entering or leaving the locality of groups of solar systems, linked together by rogue planets orbiting between them.

All neutrinos, electrons, protons, neutrons, atoms and molecules expand at the same constant rate. Although, I must point out that Mark McCutcheon remains unconvinced about the existence of the neutrino, and he believes the expansion rate of the electron is much higher than the one of the atom.

This first type of expansion of matter never changes, it does not contract either. It is gravity and it is constant, although the distance reduction accelerates. Mark McCutcheon has derived his Atomic Expansion Equation to calculate gravity in his book *The Final Theory,* and it replaces Newton's Law of Universal Gravitation, and Einstein's equations of General Relativity. Everything explained by Quantum Mechanics can easily be explained through Atomic Expansion Theory. For more detail, please read *the Debunk of Standard Theory* from Mark McCutcheon.

2.4 Everything is just expansion and contraction of orbiting and vibrating matter

How many times in the esoteric, spiritual and New Age circles have you heard that everything is just expansion? Expansion of minds, of matter, of ideas, of concepts, of worlds? Indeed, everything is expansion and contraction of matter, in five different ways. With that you can describe the entire physics of our reality and beyond.

The fourth type of expansion is the one described by Seth in Jane Roberts' books, where our reality is simply psychological or psychophysical in nature, and is projected out all around us almost instantly from a mental enclosure at the centre of our brain. After the matter of our reality has been instantly expanded out of our brain, then the matter is constantly being replaced and replenished precisely like a computer monitor does at 50 or 60 hertz, using the electrons from the main power source in our homes.

We each create our own bubble realities. Each time we think

something different, a new one is created and starts expanding. We all relatively create the same realities through an instant telepathic connection of electrical impulses, signals and codes of different frequencies. Hence, reality is equivalent to a dream or imagination or thoughts. There are different realities or timelines of different frequencies, and you move to these different timelines through thoughts and will alone, as thoughts, feelings and emotions change your vibrational frequency.

2.5 Reality is more like a virtual reality, computer simulation or holographic universe, where there is no distance

The fifth type of expansion of matter is the Theory of Universal Relativity, a theory of relative motion. People are talking more and more about the idea that we may be living in a holographic universe, a simulated world or a virtual reality. However, they have not thought further about the implications, the first one being that there is no real distance in such a reality. When you look at a simulated world on a computer screen, for example of Earth and an asteroid passing each other in space, the asteroid is actually just expanding as it approaches, and reduces in size as it goes away.

This is essentially my theory of relative motion, or Universal Relativity, which replaces Einstein's Special Relativity. Objects in the *real world* are not moving at all, as there is no space nor distance. Objects are simply expanding when they are coming towards us, and shrinking as they move away. Just as we would expect in a virtual world or a holographic universe, or a psychophysical reality.

Finally, it explains why an alien spaceship or a light being can move instantly to another galaxy or to another universe. Because there is no *real* space, there are only shrunk objects all around us. And to move there you simply have to shrink yourself to that scale and place. And as motion is relative, from the perspective of the people you are reaching, they see you enlarging towards them.

In the third type of expansion, they achieve expansion and

contraction of matter, or instant travel, simply by changing the vibrational rate of the matter composing their ship and themselves, to reach the right resonance of any specific time and place in history.

As Seth says in Jane Roberts' *The Early Sessions,* the entire universe is simply electrical in nature. Underlying all realities and all dimensions, it is all simply an electrical universe filled with signals and codes of different frequencies, that we interpret with our brain to form our camouflage realities that are just an illusion. Fundamentally, it is a canvas with a blanket of neutrinos and electrons. This is all very nice, but what does it mean exactly in real scientific terms?

2.6 What are other dimensions and planes of existence?

When they talk about different planes of existence and other dimensions, what they mean is simply matter vibrating at a different frequency or vibrational rate. Every time matter or energy (which is also always made of matter), vibrates at a different rate, then it has moved into another dimension or plane of existence. So far you might have already known that much, as it is plastered in all of the New Age literature.

What you might not know, is that all the matter composing us and the Earth is constantly expanding, as stated in the first type of expansion according to Mark McCutcheon. Hence, as we move through time, the interdimensional frequency or vibrational rate of the Earth, and all of us, is constantly getting higher, ever so slightly. Which means all the particles are moving faster and their orbit constantly enlarge. And this leads to the third type of expansion and contraction of matter, defining the vibrational rate of matter, or resonance, of every location in time in the universe.

In Dolores Cannon's *The Convoluted Universe Book Three*, a woman mentions that a being from the 18th century looking at us today, would see us as beings of light, as we would glow compared to them. And it was confirmed by Nostradamus in her other book series about him, as he described the people under hypnosis from today that came to him, as beings of light from the future. He said they were glowing. From the *Convoluted Universe:*

> "You will notice a difference in resonance, but you will already be there, because your resonance is increasing every day already, as it is. And so, all of a sudden, one day, you will reach the prerequisite cycles per second to take you from here to there. Let's explain it this way. If somebody came back right now from the eighteen hundreds to see you, you would glow to them. You've already reached those cycles per second that would glow to a human form of, say, the eighteen hundreds. So in essence, your cycles per second are raising.
>
> "Comment [by Dolores Cannon]: Could this be one reason why when John and the others went to visit Nostradamus [*Conversations With Nostradamus*], he saw them as glowing energy spirits of the future? Was this because they were actually vibrating at a faster frequency that made them glow?"
>
> Dolores Cannon, *The Convoluted Universe Book Three*

In that discussion above, cycles per second must be referring to how many times the neutrinos and the electrons are orbiting within all the particles composing us. This is what the vibrational rate or frequency of matter refers to, how fast all our particles are orbiting.

The past and the future are also other dimensions, meaning, places where matter is vibrating slightly slower or faster than in the present. Not only you can travel in time by slightly changing your vibrational rate or frequency, but you can move to other dimensions and anywhere in the universe, also by switching to the right frequency of the place and time you wish to visit. Although all of this must be easier to achieve when you are living in a world of a lesser density, or a higher dimension. This is how psychic mediums can see so much more than us.

You shrink or expand yourself to wherever you want to be, in whatever time period you want. You shrink to get somewhere, but from the point of view of your destination, you actually expand towards them in order to get there. Motion is relative, and you are not really moving. You are just expanding and reducing in size, by changing the vibrational rate of the matter composing you. But how is this achieved?

2.7 How to move to other dimensions?

When you have a solid like ice and you heat it, you are sending energy into the ice. Energy as heat, like infrared light, is made of large clusters of electrons in Atomic Expansion Theory, it is made of matter. These added particles are exciting the molecules and they become more spaced out. The result of molecules more spaced out, turns the ice into water, and eventually into vapour.

What I must stress here, is that this is all our actual science can observe and is aware of: the change of the spacing between molecules, which turns solids into liquids, and into gases. What science is not yet aware of, is what happens when instead you change the spacing within and between the atoms composing these molecules, which affects their size. And this is the key to changing your vibrational rate, your interdimensional frequency or your resonance.

There is an excellent reason why our science has never observed or thought of interdimensional density, because as soon as you change the speed at which neutrinos and electrons are orbiting within atoms, between atoms, and between molecules, you also change the size of the atoms and of the molecules, as well as the spacing between them.

At that point, matter moves outside our range of vision, outside of visible light, and light and any normal matter pass right through it, because it is moving in fast forward compared to us. Or at the very least, it all intermingles together while occupying the same space, like solids existing through water or gases, which are all states of matter of different densities.

This is what defines other dimensions, normal matter made of smaller or larger atoms and molecules, due to the electron's growth within the atoms, creating matter that we cannot see, feel or interact with. Most probably because when matter vibrates that fast, or that slow, it is too fast or too slow for our range of vision or our instruments to register. And they probably don't need to vibrate that much faster or slower, before the change happens.

Each dimension still has its own first type of density of matter, where solids can turn into liquids and gases, and vice versa. In this case, the electrons, atoms and molecules get more excited, but they don't grow or shrink in size.

This creates different realities within realities, physical in nature, but psychological in origin, since everything is just a construction of our brain to begin with. Our brain will only construct something for which it receives information about, through signals of different frequencies. We would need to develop specific technology to access these other dimensions. And a good understanding of how it works, and an awareness that it exists, would appear to be the first step.

It is very important to understand here that a frequency, which defines the vibrational rate of matter when it comes to other dimensions, refers to the speed at which the neutrinos are orbiting the nuclei of electrons, and the speed they are orbiting *between* the electrons, enlarging or shrinking the size of their orbits, and consequently the size of the electrons. In turn, the electrons do the same within the atoms and between the atoms, affecting the size of the atoms and the molecules, as well as the spacing between them.

Raising your vibrational rate, is raising the speed of your neutrinos and electrons, which in turn makes the atoms and the molecules slightly larger, as well as enlarging the spacing between them. This is the third type of expansion and contraction of matter, more specifically defined as growth and shrinkage of matter.

This changes the density of matter, in a different way than is defined in our physics today, and this is why a fourth dimensional world is also called a fourth density world. The definition of dimension here, refers to the size of the particles, and consequently of objects and beings. It is unrelated to our three dimensions of length, width and height, to which could be added a fourth dimension, such as Einstein's fourth dimension of time. Neither is it related to geometrical shapes of higher mathematical dimensions, such as hyperspheres and tesseracts.

In a fourth dimensional world, everything is exactly as on Earth, it is a physical world just like here. But the density of the matter is lighter, all the particles are larger and more spaced out, and probably more malleable too. All the particles are moving, orbiting and rotating faster, and so is the way beings in the fourth density world live, at a faster time rate.

Many have said that several planets of the solar system are inhabited by beings living in higher dimensions. If you consider Jupiter and Saturn, which are said to be fourth dimensional, from our third dimensional perspective they appear as gaseous planets. However, for the civilisations living there on these planets, who are fourth dimensional already, their planets would appear as physical, as solid and as normal as Earth appears to us, their planet would not appear gaseous to them.

We might be surprised when we visit these planets, because what we perceive as gases, might actually be matter made of larger atoms and molecules than the ones we find on Earth, with matter vibrating faster. Essentially different types of larger molecules with larger atoms, instead of molecules the same size as on Earth and simply more spaced out.

As we are said to be multidimensional beings, and dreaming is already like living in a higher dimension, we possibly have another body superimposed over our physical one, made of particles orbiting faster, with which consciousness could continue to exist after death. The esoteric literature has long talked about our etheric bodies.

This is also, most probably, what the bodies of people from other dimensions and times, such as spirits, ghosts, ultraterrestrials, the Bigfoot and the Mothman, are made of. They are made of real physical matter, atoms and molecules, made of particles either smaller or larger than ours, and more or less spaced out, depending on which dimensional or density plane they exist on. Larger if they are in a higher dimension or density, smaller if they are in a lower dimension or density. This is a second type of density of matter, called interdimensional density of matter, not to be confused with our normal type of density.

If they have red eyes looking more like lights, like it has been reported for the Bigfoot and the Mothman, then they probably come from a lower density, and they see the infrared light part of the spectrum instead of visible light. In higher dimensions, many species of aliens might instead see the ultraviolet part of the spectrum.

But remember, that they each have a slightly amended electromagnetic spectrum compared to ours, with clusters of electrons made of larger or smaller particles, either moving faster or slower, therefore clusters of different sizes, and more or less spaced out than ours. Therefore, using infrared or ultraviolet filters to see them might not always work, although many who have tried have been successful. Their reality cannot be that far removed from ours, or we would not encounter them so often.

To appear to us, they only need to either accelerate or lower the frequency at which their electrons are orbiting their particles, which would shrink or enlarge their orbit. And of course, sometimes natural phenomena achieve just that, and we see them in our reality. Either because of power plants producing a lot of heat, or sources of high magnetic and geomagnetic fields, perhaps even sounds we cannot hear but that dogs and cats might pick up. Sometimes people end up in their reality, and sometimes they don't come back. Which would account for many of these mysterious disappearances every year, which are reaching alarming numbers.

There could be small civilisations of humans living in other dimensional worlds, and perhaps not even all in the same one. I would think they should have no problem surviving there. It would just be like a different reality, where there would not be eight billion of us, and probably populated by other beings of some kind. Vegetation and animal life might also be different, and either be much larger or much smaller depending on the density of that world, which would affect gravity there.

As I mentioned earlier, in time the vibrational rate of matter constantly rises, with all particles orbiting faster, with their orbit also enlarging faster, causing a constant expansion of matter. So much so that we would look like glowing beings to people from the 1700s, who might easily believe we were angels or demons. Two hundred years is not a very long period of time, so this effect must be noticeable, even measurable. There must then be some kind of proof to this, and there is.

All our matter and the one of the Earth is becoming less dense as time goes by, moving us into higher dimensions. Which means that, at the time of the dinosaurs, 65 million years ago, the world must have been extremely dense. Remember, this is a different kind of density than the one measured today, which turns solid into liquids and gases.

We usually associate denser objects with heaviness, and today denser objects indeed weigh more than normal ones of the same size, since density affects gravity on the ground. However, a difference in density for objects of the same size, would not affect falling objects. They would fall at the same rate, since in this case, only the size of the objects matters when calculating gravity. While for the UFOs floating in the air, this is buoyancy of an object of a lesser density, being pushed above the air, which is denser or about the same as the UFOs. Like a rising helium balloon, which is less dense than the air.

So how did massive dinosaurs roamed the Earth and flew around so easily, when scientists are telling us that they appear to have defied the laws of physics? And these scientists are not even aware of the fact that the world must have been denser by a large margin back then.

Usually, smaller animals and insects tend to do everything faster than larger ones, who in turn tend to be denser and move in slow motion. In those days, matter could have been denser, and yet gravity much less than it is today. But how?

As time goes by, the Earth becomes less dense, which means it enlarges and grows in real time. A larger object leads to more gravity, according to the Atomic Expansion Equation, where gravity depends on the size of the objects.

This said, if the density of our body is also becoming less dense in time, then we should feel gravity much less, as we would weigh less. But perhaps not, if everything else around us becomes less dense as well, at the same rate.

At the time of the dinosaurs, the Earth must have been denser, but smaller than it is today, leading these dinosaurs to weigh much less than we initially thought. This is not due to the difference in density compared with today, but to a lesser force of gravity that comes with a smaller planet. Less gravity also leads to larger vegetation and animals.

That the Earth is growing in real time, leading to a constant rise in gravity, seems to be real, if you check out the videos of Neal Adams on his website *www.nealadams.com/science-videos.* Particularly his video on YouTube called *"Neal Adams - Science: Part 1 - Conspiracy - The Earth is Growing" (https://youtu.be/oJfBSc6e7QQ)*. Here is a convincing argument that at one time the Earth was very small, and all the continents glued together covered the entire planet. There were no oceans back then, and the continental drift every year is proof that Earth continues to grow even today.

In time the Earth is becoming less dense, but not in the way of a solid turning into water, and more in the way of moving into a higher dimension. All its particles become more distanced and grow in size, and this leads to a tangible and concrete growth of the planet, and to more gravity.

This falls into the third type of growth and shrinkage of matter, involving the neutrinos and electrons moving faster as time goes by. This causes an enlargement of the atoms and molecules and all of the orbits, moving the Earth into a higher dimensional world, or higher density world. It could be a proof of this third type of expansion of matter, which would lead to a real and concrete growth of matter as the density lessen.

2.8 How to raise the vibrational rate or interdimensional frequency of matter?

Although you would use heat or pressure in order to turn solids into liquids and gases, so the molecules become more spaced out and volatile, it is not the best way to raise your vibrational rate and move to another dimension, although it might work under certain circumstances. The process is different when you need to cause the electrons to grow inside and in between the atoms, as well as being accelerated, so matter can grow while still keeping a coherent and solid molecular structure.

Ultraterrestrials appear to have the technology to do that, as other beings appear to be able to achieve it through will. Some sources are telling us we can achieve it through thought alone, by imagining white light vibrating faster passing through us. However, there must be a way to achieve it through technology.

I feel that light composed of clusters of electrons, with electrons orbiting faster within the clusters and between the clusters, is one way to raise the vibrational rate of matter. The difference from our normal light, is that the electrons are larger, the speed and size of the orbits of the electrons are higher, defining the size of the clusters of electrons within light, and the distance between them. By exposing an object to that higher light frequency, you would raise the frequency of that object. But how can we produce such interdimensional light, from a higher electromagnetic spectrum, where all the particles are larger and vibrating faster?

Please note that the frequency of light I am referring to here is not related to the frequency of light we find solely within our own electromagnetic spectrum, where light goes from infrared to visible light, to ultraviolet light and up to gamma rays. That frequency of light simply means a change in the size of the electron clusters composing light or radiation, the electrons themselves don't change size or speed. And the intensity of the light, is how many clusters are passing per second. The clusters being larger for infrared, and very small for gamma rays, which means with more or less electrons within the clusters. The interdimensional frequency of matter, refers to the number of orbiting cycles of neutrinos within and between electrons, and of electrons within and between atoms and molecules.

I will soon discuss my new model of the electron, which now has neutrinos orbiting around a nucleus called a super-neutrino. The speed of the neutrinos affects the size of their orbits, which causes the electrons to grow or shrink. This in turn affects the speed and the size of the electrons, and the range of their orbits. Therefore, the frequency of light I am referring to, is the changing speed and size of the neutrinos' orbits within electrons, between electrons, and between the clusters of electrons.

Technically speaking, every dimension has its own corresponding adapted electromagnetic spectrum, where the clusters of electrons within the light are composed of electrons of varying sizes and speeds, and of varying orbit sizes, changing the size and spacing of the larger particles they compose, such as the clusters of electrons within light.

Another way of raising our vibrational rate would be through sounds of higher frequencies, as the displacement of matter and the higher vibrations, could raise the speed of the neutrinos and of the electrons within and between the atoms and the molecules, until they reach the desired resonance.

However, the most likely way we could develop technology to accelerate the rate of orbiting electrons within atoms, and between atoms, would be through powerful rotating electromagnets, such as the ones Ralph Ring along with Otis T. Carr have built to levitate flying saucers. You can search online and on YouTube for Ralph Ring, here are the main pages to start your investigation:

- http://projectcamelot.org/ralph_ring.html
- http://bluestarenterprise.com

> "In essence, the account is as follows. Carr and a small group of engineers and technicians, one of whom was Ralph, built a flying disk, powered by rotating electromagnets in conjunction with a number of small, ingenious capacitor-like devices called "Utrons". A number of prototypes were built, ranging in size from experimental models a few feet across to a passenger-carrying craft which was fully 45 feet in diameter. The smaller disks flew successfully - one even disappeared completely and was permanently lost - and Ralph himself testifies to having co-piloted, with two others, the large craft a distance of some ten miles, traversing this distance instantaneously."
>
> Ralph Ring

I have not yet fully investigated Ralph Ring and Otis T. Carr, but from a quick overview, their devices have changed the molecular structure of matter:

> "When recalling the heady events of the late 1950s working day and night with Carr, Ring again and again stressed that the key was working with nature. "Resonance", he would emphasize repeatedly. "You have to work with nature, not against her." He

described how when the model disks were powered up and reached a particular rotational speed, "...the metal turned to Jell-o. You could push your finger right into it. It ceased to be solid. It turned into another form of matter, which was as if it was not entirely here in this reality. That's the only way I can attempt to describe it. It was uncanny, one of the weirdest sensations I've ever felt.""

Ralph Ring

These statements prove, more or less, what I was explaining above. By accelerating the orbiting electrons within the atoms, the atoms become larger and more spaced out, creating a new type of molecule at a higher dimension. The ship that disappeared completely might still be there, you might see it through the use of ultraviolet light (higher dimension) or infrared light (lower dimension), or colour filters. It is in a higher dimension, which means a physical world of another density of matter that we cannot see or interact with. The metal turned to jelly, matter of a lesser density, probably as light as air. This is how large stones such as Stonehenge, the Pyramids, and the Coral Castle of Edward Leedskalnin in Florida, were easily made, shaped and transported to where they were needed.

Ralph Ring and his team also apparently moved one of their ships ten miles away instantly, and back, through teleportation, meaning to the resonance or interdimensional frequency of matter at that precise location and time. To travel in time, or instantaneously to the other side of the galaxy, you only need the specific resonance or frequency of the matter there at that specific moment in time. Even just moving ten kilometres away, within the same dimensional world, is moving to another dimension within the same overall dimensional world, where the vibrational rate of matter in every nearby location is relatively similar, but not exactly the same.

Since matter is constantly expanding, the electrons within the atoms are constantly enlarging their orbits and moving faster, and the frequency of matter in any place and time is continually getting higher, and is unique. There are as many different dimensions or planes of existence as there are different rates at which the electrons are orbiting, hence it is infinite. Eventually the Earth and everything on it will move into what could be considered the fourth dimension. The interdimensional frequency of our matter might have already crossed that threshold, although we would not have noticed anything different.

The concept of the third, fourth and fifth dimensional worlds or densities of matter, can only be arbitrarily defined by convention, as particular densities of matter within which several stable worlds and civilisations of a particular vibrational rate can exist and thrive. All ultraterrestrials, or aliens, are said to come from higher dimensions, which explains why most of the time we don't see them. But apparently, they have always been here, they are everywhere, and they have their own cities.

It is worth noting that Ralph's team, that teleported 10 miles away instantly, were still aware and doing things after the move. But once they came back to a normal vibrational rate, they could not remember anything. They had to rely on other evidence to confirm what they did during that period of time, when they were vibrating faster in a higher dimension:

> "Did the craft fly? "Fly is not the right word. It traversed distance. It seemed to take no time. I was with two other engineers when we piloted the 45' craft about ten miles. I thought it hadn't moved - I thought it had failed. I was completely astonished when we realized that we had returned with samples of rocks and plants from our destination. It was a dramatic success. It was more like a kind of teleportation.""
>
> Ralph Ring

The ultraterrestrials, or aliens, don't need to wipe out our memory for us to forget, unless we are under hypnosis to recall everything from our subconscious mind. It is simply that memories made in higher dimensions or planes of existence, are only remembered with great difficulties once we move back to a lower dimension or vibrational rate of matter. It might be a limitation of our physical brain.

Precisely like in dreams, which are apparently realities made of atoms and molecules as well, according to Seth in Jane Roberts' books. Dreams are also psychophysical realities identical in every way to our own physical world, where we need to immediately set ourselves to remember our dreams once we awake, before it is all gone forever.

Dreams are just another dimension, like any specific space and time anywhere in the universe is another and precise dimension where matter vibrates at a specific interdimensional frequency or resonance. We could all move into or see the dream of another person, by simply vibrating at the same frequency of matter within that dream.

I have experienced it, I could see the dreams of one of my friends while I lived in Los Angeles, he's a psychic medium. He also had the ability to come into my dreams where we had conversations, which we confirmed the next day over the phone. His dreams were of a quality and clarity that were even more real than reality. I also could not lie to him, he always knew when I did, it was extremely annoying. It is how telepathy works and how people sometimes share dreams, or the same collective hallucination. A hallucination is therefore a real physical or psychophysical reality, or what I like to call another bubble reality.

To get back to Ralph Ring, he also stated:

> "What's more, time was distorted somehow. We felt we were in the craft about fifteen or twenty minutes. We were told afterwards that

we'd been carefully timed as having been in the craft no longer than three or four minutes."

Ralph Ring

The faster your electrons are orbiting the nuclei of atoms and between atoms, the more you can do within a same amount of time as someone not vibrating as fast. There is still then a relativity to time, in its perception, called interdimensional time. I must point out that this has nothing to do with how Einstein posited the relativity of time, that he thought could be influenced by gravity or simple acceleration. Einstein's theories are incompatible with Atomic Expansion Theory.

One last statement from Ring:

"The Utron was the key to it all. Carr said it accumulated energy because of its shape, and focused it, and also responded to our conscious intentions. When we operated the machine, we didn't work any controls. We went into a kind of meditative state and all three of us focused our intentions on the effect we wanted to achieve. It sounds ridiculous, I know. But that's what we did, and that's what worked. Carr had tapped into some principle which is not understood, in which consciousness melds with engineering to create an effect. You can't write that into equations. I have no idea how he knew it would work. But it did."

Ralph Ring

How can we explain that? Thoughts and will can affect our own vibrational rate, which once amplified by their rotating electromagnets, and specifically shaped capacitors (Utrons), was extended to the ship they were on. Aliens often say they only have to think of where they want to go, and they are instantly there.

This is quite useful, because we appear to instinctively know what is the vibrational rate of a certain place and time where we want to go, without having to know for sure what is the exact frequency of matter there. After all, this is a psychological or psychophysical reality right out from our own brain. However, I am sure eventually technology could measure, calculate and extrapolate such interdimensional frequencies or resonance of different places and times, and thoughts would then not be required to travel there.

2.9 How can UFOs fly so easily and how to build an anti-gravity device?

We can see how easily spaceships can fly through the air using the geomagnetic fields on Earth as roads, and other astronomical magnetic fields populating space when in outer space. By accelerating the speed or frequency of their electrons within and between atoms, their particles become larger and more spaced out, and they become less dense. Anything that is less dense will automatically be pushed up above what is denser. And now we know the secret of these flying Tibetan Yogis, and it can all be explained by physics.

A balloon underwater will immediately float to the surface, because the air inside it is less dense than the water. And if there is helium in it, which is less dense than the gases in our atmosphere, it will fly up into the sky. As soon as a spaceship becomes less dense, it could immediately fly off out of the atmosphere without any need for propulsion, just through what is less dense being pushed up by what is denser.

Indeed, they must then use magnetic fields to stay on Earth and navigate, otherwise, they are so light, they would be expelled instantly out of the atmosphere. Or they can easily control their exact vibrational rate of matter, in order to be in the perfect density, to float and fly wherever they want.

It might also be possible that a less dense object would be pushed out of the solar system, but I can't be sure. The gas giants

Jupiter and Saturn, which are planets of a lesser density, are at the edge of the solar system, while denser planets are nearer the Sun. Just like it is at the bottom of the ocean, where the pressure and the density of the water is more significant.

The density of matter populating our solar system appears to be denser nearer the sun. Which would be logical from the formation of the solar system, following an initial massive explosion. The expansion would gather more matter nearer the centre and be denser, while the disks of matter orbiting further away, that would go on to form the outer planets, would be less dense.

But space itself is rather empty, density wise. And yet, the speed of a planet around the Sun defines the distance of its orbital ring within the solar system. Just like the speed of a satellite defines the height of its orbit around the Earth, and how fast it will go around the planet. But perhaps it has less to do with the density of matter, and more with the geometry of expansion.

There is no such thing as an ether required in Atomic Expansion Theory. Space is filled with electrons everywhere, in the form of clouds for magnetic fields, or clusters for light, heat and other radiant forms of energy. No medium is required in empty space for them to move through the relative motion of the geometry of expansion.

Strong magnetic fields might be another way to raise the vibrational rate or frequency of matter, changing the density of matter so it becomes extremely lightweight. At this point, from the point of view of an object vibrating faster, everything is lighter and going faster. In a same amount of time, you can achieve much more, as you are essentially moving into another dimension, changing the size of molecules, atoms and electrons, and the spacing between them.

Several spaceships appear to be using crystals in order to amplify the background radiation (gamma rays) and/or light surrounding us everywhere, to produce or direct energy, and use it as a mean of propulsion, especially in space. On Earth, some of these spaceships appear to have all around them a magnet turning one way, and another inside turning the other way, essentially creating huge electromagnetic fields that they can use to navigate on Earth, using the geomagnetic fields.

What we did not realise before, was how easy it is to propel such ships when their density is so light, that they are lighter than the gases in the air. It takes almost no energy at all. And the momentum of light through kinetic energy can easily be used to propel such ships very rapidly over extremely long distances. They are not in our dimension, although sometimes we can see them. They are much lighter, less dense. The only way we can reproduce that technology would be for us to build ships and technology capable of vibrating faster.

2.10 The infinitely small is identical to the infinitely large universe - New models for the atoms as galaxies, electrons as solar systems, super-neutrinos as suns, neutrinos as planets, and sub-neutrinos as moons

In a world completely defined by the expansion and contraction of matter, you would think that there would be no difference between the microscopic world of the infinitely small, particle physics, and the macroscopic world of the very large, astronomy. What we see in the subatomic and atomic world must essentially be identical to what we observe in the night sky. It might be thought-provoking, especially in this day and age of advanced physics, but it all works perfectly well.

For a long time, scientists were trying to compare the atom to a solar system, it was essentially originally the model of the atom according to Ernest Rutherford and Niels Bohr. It did not quite work as expected, and so the idea was abandoned. What I am saying instead is, it is the electron which is identical to a solar system.

I thought of it because, the way the atoms are bonding together to form the molecules, in attracting and repelling forces, is through the exchange of electrons between them. It made me wonder how electrons could bond together to form larger stable particles, such as protons and neutrons within atoms, and clusters of electrons within light and heat, if the electrons themselves were not also exchanging even smaller particles between them, which are called neutrinos. Especially when, in Atomic Expansion Theory, there are no more electric charges, and everything must be explained by the expansion, motion and behaviour of particles between each other.

Neutrinos at the scale of our very large are planets, while sub-neutrinos are moons and super-neutrinos are suns. Anything smaller than moons, such as asteroids, comets, rocks and grains of sand, are sub-sub-neutrinos. This is how much they can vary in size, and these are all different names sub-sub-neutrinos get at our larger scale.

It could be debated that these are all different types of particles, and we could each give them a name instead of just calling them all sub-sub-neutrinos, no matter the size. And indeed, those names are fine: asteroids, comets, rocks and grains of sand, then, exist at all scale universes. For convenience, I call them all sub-sub-neutrinos of varying sizes in our very small.

Sub-sub-neutrinos would be the smallest particle in existence at the scale of our very small, although they would be made of another set of particles even smaller at a much smaller scale than our very small, from another scale universe. All particles are made of sub-sub-neutrinos, meaning that if you smash any particle, this is what you will end up with. Which is why in beta-decay (radiation), neutrons decay into one proton, one electron and one or several neutrinos of varying sizes. This is essentially what the weak nuclear force is, one of the main four forces of nature.

Our computers, eventually, will easily be able to build realistic virtual 3D realities, of any dimensional world, because it is just a question of programming. One only needs to instruct the electrons to form into protons, neutrons and atoms, and then into molecules, in order to obtain a physical world as real as the one we are living in. At this time, virtual realities are entirely made of electrons, without bothering with assembling them into atoms and molecules, and without writing our real laws of physics into the programme, to get them to form concrete 3D objects through such laws.

If electrons are solar systems, then atoms are galaxies. This is not going to be easy to get people to accept this idea, because this is not how we visualise atoms in our science at the moment. Atoms don't appear flat for a start, while most galaxies are. But atoms move so fast from our perspective, we think they are spheres, when in fact atoms must be mostly flat objects, just like galaxies. After my discovery, I searched the Internet for some confirmation, and I believe I read in *The Urantia Book,* that the atoms are flat objects which only appear to be spherical. So that is at least one confirmation from the esoteric literature.

People never thought electrons could be more than just small balls or point particles, without any real size or mass, or that they could have a nucleus with smaller particles orbiting within and between them. In my book *New Age Physics,* I make the case for this in more detail, in chapter 4. But suffice it to say, an atom could easily resemble a galaxy.

Several solar systems within the arms of galaxies would be linked together through what is called rogue planets, meaning planets orbiting between these solar systems, but far away from their nuclei. Which means very far away from our eight known planets. If people believe there is only one more planet to be found, planet X, they're up for a surprise.

Sirius and the Pleiades for example must be linked to us. There must be rogue planets orbiting around these star systems and ours, and other planets orbiting only our star but far away, nearer the midpoint between us and these solar systems. Possibly these planets can switch their orbit from one sun to another in an 8 shaped orbit. There could be countless planets everywhere between and even beyond these solar systems and ours, linking them together. We have not yet seen these planets, bonding solar systems together into larger stable particles at a higher scale, similar to what we would call protons and neutrons within the nucleus of an atom.

The entire visible galaxy would only be the nucleus, as rogue solar systems orbiting between galaxies, at a far distance from the nuclei or visible galaxies, must be linking them together to form molecules at a much larger scale. These cosmic molecules would also be linked together through orbiting rogue solar systems between them.

Hence, solar systems are linked together by rogue planets to form larger stable particles such as protons, neutrons and clusters in light and radiant forms of energy. And galaxies also bond together through rogue solar systems to form cosmic molecules, that most probably compose a much larger body of something at a larger scale.

Who knows, the entire visible universe might be forming a being of some sort, maybe what we came to call God. But you can understand the scale of it, if our milky way galaxy is only one atom of this entire being, while we would be living on a neutrino within one electron of that being. We are small indeed, but of course it is all relative, when you consider that there is no space or distance, in a universe where matter simply expands and shrinks constantly instead of moving, if you followed my initial discussion.

Reality would just be a mental construction of a physical nature, just like dreams and thoughts are. The difference is how we perceive it, as reality is perceived via our external senses, while dreams and thoughts are perceived through our internal senses. All these planes of existence, like other dimensions, are made of matter and are physical in nature, as Seth would say in Jane Roberts' *The Early Sessions*. All these realities are made of electrons, atoms and molecules, but of different sizes and vibrational rates or frequencies.

2.11 Gravity - No speed of light limit, the speed of light is relative and variable even in a vacuum

In Atomic Expansion Theory, there is no more Einstein or concept of space linked to time or to mass-energy, hence there is no more speed of light limit in our universe.

The speed of light is not constant, nor is it constant for all observers in different locations going at different speeds. It is not time that is relative, it is the speed of light, which is limited by its speed, creating optical illusions.

The electrons composing light, in the form of electron clusters, are simply expanding freely as soon as they are released at high speed from the subatomic world. Mark McCutcheon calls this the crossover effect, when electrons are moving out of the subatomic and atomic world, which is also the second type of expansion and contraction of matter, more specifically called growth and shrinkage of matter.

The difference with the third type discussed in the previous sections, is that these growing and shrinking electrons are externalised, they are outside the atomic realm, so they don't affect the size of the atoms and the molecules, or the spacing between them.

As McCutcheon says, radio waves through to microwaves are typically produced with oscillators - either wire antennas where electricity oscillates back and forth, or hollow tubes designed to produce resonating or oscillating electromagnetic waves, or bands of compressed electrons. While infrared radiation through to gamma rays, including visible light, are essentially produced by heating a material until it "radiates energy" into clusters of electrons. This difference splits the electromagnetic spectrum into two specific and distinctive phenomena.

In order for clusters within light to be freely launched into space, the neutrinos orbiting electrons, and between electrons, must reach a certain speed, and their orbits a certain size (the crossover effect).

Depending on the size of the clusters, they are launched at sensibly the same high speed, which accounts for most of the speed of light. Once they have been launched at that certain speed, the initial boost to free themselves from the subatomic world, the speed of light is further calculated between objects using the Atomic Expansion Equation of Mark McCutcheon.

I have assessed that the expansion rate of the electrons is the same as the one of the atoms, although for Mark McCutcheon the expansion rate of the electron is unknown, and radically faster than the expansion rate of the atom. As a consequence, to calculate the changing distance between expanding subatomic particles, we can equally use the Atomic Expansion Equation, so I changed its name to the Atomic and Subatomic Expansion Equation.

2.11a Atomic and Subatomic Expansion Equation to calculate gravity between expanding objects made of atoms and/or electrons

$$\mathbf{D' = D - n^2 X_A x (R_1 + R_2) / (1+ n^2X_A)}$$

D' is the changing distance between two expanding objects of radius $\mathbf{R_1}$ and $\mathbf{R_2}$

D is the original distance between the two objects

n is the number of seconds that have passed since the original distance was measured between the two objects

$\mathbf{X_A}$ is the universal atomic and subatomic expansion rate's constant 0.00000077 $/s^2$ or 7.7×10^{-7} $/s^2$

The top of the second part of the equation refers to the absolute decrease in distance due to the constant internal expansion of the objects, and the bottom part refers to the further relative decrease in distance, since space has not expanded along with the objects. That bottom part is what causes the acceleration in the changing distance. For the full explanation, see chapter 3 in *New Age Physics,* section *3.5 Two types of distance decrease to calculate gravity*. This Atomic and Subatomic Expansion Equation to calculate gravity replaces Newton's Law of Universal Gravitation and General Relativity's equations.

Thus, the speed of light, adding to its initial launching speed, will depend on the distance between the origin and the destination, and on the size and the expansion rate of these objects. One of these objects being the electron clusters composing the light, the other object being the destination, for example a planet or a spaceship.

One must also consider the orbits of the electron clusters in light, since their path will deviate as they pass near larger objects such as the Sun, following the natural orbit effect of the geometry of expansion. These partial orbits would further accelerate the speed of light, just like satellites sent into partial orbits around planets, are accelerated through the slingshot effect, which is entirely a geometrical effect of relative motion, involving no physical force of any kind.

And then, the clusters could be slowed down by whatever they encounter along the way, such as gases (molecules and atoms) and magnetic fields (clouds of electrons). The speed of any light coming to us must be decelerating and accelerating alternatively, whether we accept that the expansion rate of the electron is the same as the one of the atom, or not. Light is not immune to gravity, orbits, and the geometry of expansion. The speed of light varies, it slows down with anything it encounters in space, and accelerates with gravity and the slingshot effect, when passing by large bodies.

This said, light, just like heat and radio signals, tends to dissipate in time with distance, and become weaker. The clusters of electrons within light and heat, and the compressed bands of electrons in radio signals, eventually lose coherence and disperse into space, which would also affect its speed.

In order for the electrons to leave the atom in the crossover effect, they must be excited, and their neutrinos must be orbiting faster. Like when we are applying heat, which is a way to add particles and clusters to a system, in order to excite all the other particles. At this point, the orbits of the neutrinos enlarge and the electrons grow. They may also shrink if the neutrinos suddenly slow down, which happens as they come into contact with other smaller electrons or clouds of electrons (magnetic fields), with slower orbiting neutrinos.

Each time the clusters of electrons in light return to the subatomic realm, such as when passing through a block of glass, they slow down. But the ones that will escape are re-launched at high speed, as once again they have reached the speed required to leave the subatomic realm. And no doubt they are helped by the accumulated momentum since leaving the original light source, which some of it must be kept as they go through the glass without hitting any of the atoms and molecules. And the collisions might cause some atoms to collapse or explode, freeing more electrons and clusters at high speed.

This is essentially the crossover effect of Mark McCutcheon described in *The Final Theory*, which I called the second type of growth and shrinkage of matter. It explains the attracting and repelling forces of the electrons in static electricity, as when two clouds of expanding electrons meet between two rubbed rods made of glass or plastic. If the electrons within the meeting clouds of electrons, between the rods, are the same size, they will repel. If they are not, they will equalise and attract. Or they will neutralise if the there is an electric discharge, and the electrons dissipate everywhere, or return to the subatomic realm of the rods. Static electricity is covered in detail in chapter 3.

These electrons, electron clusters, electron clouds, or compressed bands of electrons, must all be ejected more or less at the same speed, when freed from the subatomic realm. However, once freed, their speed is affected by gravity, since they still have a constant inner expansion, like everything else, and gravity is the changing distance between expanding objects.

Hence, the speed of light is not constant, it varies greatly from one location to another, or depending on an observer's speed and direction of motion. A spaceship moving at twice the speed of light, would still see light coming to them. Because it is a question of the clusters of electrons of light, simply moving from the source of the light, to their ship in space, and it does not matter what speed they are going at when the clusters arrive.

And if they turn on a light on their ship, that source of light, and the ejected electron clusters, are already going at twice the speed of light along with them, so it is not an issue for that light to move from the source to their eyes. Although they would measure a normal speed of light in this case, on Earth we would measure that speed of light as going at three times the speed of light. The speed of light is relative to the point of view.

Going at twice or more our actual measurement of the speed of light, is now possible in Atomic Expansion Theory, if the acceleration is gradual and constant, in order to avoid being crushed, and if we could find a suitable mean of propulsion.

I must mention that Mark McCutcheon does not agree with me on certain of these points, although he stated that he does not believe that the speed of light is constant. For him, the speed of light is dependent on the greater and constant inner expansion rate of the electrons and of the subatomic particles, compared to the lower expansion rate of the atoms, and this explains how they can escape the atoms. But he certainly believes we can go faster than the speed of light, stating that there is nothing preventing us from trying right now, it is just that we never tried yet.

2.12 Orbits - Newton's Law of Universal Gravitation and Coulomb's Law are similar, but they measure different phenomena

The Earth and the Sun are constantly expanding, and the distance between them is reducing in an acceleration, always. So why are they not coming closer together and hitting each other?

Behind the scenes, where we can see the expansion, if two expanding objects meet in space without colliding, and they are going at the right speed to continue on their course away from each other, despite their inner expansion, then in our resulting reality, where we cannot see the expansion, they follow the natural orbit effect of the geometry of expansion, and could continue on their orbit forever.

The question now becomes, although their expansion is constant, if the distance reduction between them is accelerating, as it is in gravity, how can they continue to escape each other forever?

At that moment, when they enter an orbit, and moment by moment thereafter, the relative motion of the geometry of expansion, constantly accelerates them away from each other. This is like the slingshot effect, when a space probe is sent into a partial orbit around a planet to significantly increase its speed. It is all due to the geometry of expansion, no force is involved.

The geometry of expansion will result in a distance reduction between two objects until they collide, or, to an orbit where they may never collide. Behind the scenes, we can see the expansion, objects don't enter into orbits, and they might be moving in straight lines or slightly curved paths, away from each other. The difference behind the scenes, is that the objects are coming closer together and will collide as they expand, or, they are moving away from each other, possibly in straight lines, fast enough to escape their constant inner expansion, always keeping a certain distance between them.

In one case it is like an object falling back to Earth, in the other it is like an object being accelerated to a certain height over the Earth, where the geometry will naturally start accelerating it away from the planet and keep it in orbit. Perhaps even always at the very same location, as in a geosynchronous orbit.

For example, in an elliptical orbit, a planet is moving away from the Sun fast enough to escape their constant mutual inner expansion. Eventually the distance between the planet and the Sun reduces, as the expansion of the Sun catches up with the planet, and then the planet slows down. The planet starts to accelerate back towards the Sun, because in gravity the distance reduction is an acceleration. It then goes around the Sun and moves away accelerated, just like in the slingshot effect. Then it goes for another orbit away from the Sun, where eventually the expansion of the Sun will catch up with it once again.

The orbit may appear very stable and always relatively the same to us, but behind the scenes it is actually an accelerating orbit enlarging in a spiral, counteracting their mutual constant expansion, and counteracting the acceleration of the distance reduction between them. I visually cover orbits in more detail in chapter 3.

You can see here how the distance reduction in gravity is causing the elliptical orbit, and how orbits are linked to gravity, and how it all results from the relative motion of the geometry of expansion. But as the objects are always expanding at a constant rate behind the scenes, no matter what, there is always an overall reduction in distance between everything, even when everything is kept at the same relative distance because of the enlarging orbits.

The planets' orbital rings around the Sun are enlarging, as the entire solar system is also expanding constantly. We would see the distance between two expanding solar systems reduce in an acceleration, however they too are kept apart by rogue planets orbiting between them. The distance reduction is real in any case, gravity still happens, even if from our point of view, the distances all appear to remain as they are. With this in mind, let's look at the difference between Newton and Coulomb's equations.

2.12a Newton's Law of Universal Gravitation

$$\mathbf{F = G \times m_1\, m_2 \,/\, r^2}$$

F is the attracting force between two atomic objects of mass $\mathbf{m_1}$ and $\mathbf{m_2}$
r is the distance between the centre of the masses
G is the gravitational constant (6.674×10^{-11} N · $(m/kg)^2$)

It so happens that this equation is very similar to Coulomb's Law in static electricity.

2.12b Coulomb's Law

$$\mathbf{F = k_e \times q_1\, q_2 \,/\, r^2}$$

F is the attracting or repelling force between two charges $\mathbf{q_1}$ and $\mathbf{q_2}$
r is the distance between the charges
$\mathbf{k_e}$ is the Coulomb's constant ($8.9875517873681764\times10^9$ N m^2 C^{-2})

Gravity is the changing distance between two or more expanding objects, whether these objects are made of atoms and molecules, like Earth, or whether they are made of electrons or clusters of electrons, like light and heat, or whether they are made of clouds of electrons as in electric and magnetic fields. This is what has been referred to as the attracting force acting at a distance in Newton, and this is what his equation mentioned above measures.

This attracting force also exists in Coulomb in electrostatic, since everything constantly expands, reducing the distance between everything. However, the resulting distance reduction is counteracted by the constant enlarging orbits of neutrinos within electrons, and the constant exchange of neutrinos between electrons, keeping them bonded but apart.

In a static field, this expansion of all orbits is unseen, because everything expands relatively to each other at the same constant rate. This is identical to what I was mentioning earlier, concerning the orbit of the Earth around the Sun, and the orbits of rogue planets around solar systems, keeping them all apart despite the distance reduction.

Coulomb's Law measures this force required by the orbits to keep the objects apart, which depends on the charges of the particles. Coulomb's Law works mostly in static electrical and magnetic fields, but they are never really static, if everything behind the scenes is constantly expanding, including all the orbits of all these particles.

At the base of a static field, in Coulomb, there is a force counteracting gravity and its resulting reduction in distance between everything (Newton). So there is a link between these two equations, the same link discussed earlier where two objects will hit each other or enter into an orbit instead. But why then is there an attractive and repulsive force in Coulomb?

Coulomb uses the energetic content of the subatomic objects in his equation, while Newton uses the mass of the atomic objects from centre to centre. The Atomic Expansion Equation for gravity, mentioned in the previous section, uses instead the size of the objects, and their expansion rate from surface to surface, this is why the radii of the objects are used in the equation.

There is bound to be some equivalence when using these equations, since the energy content, the mass and the size of an object are usually proportional. The larger an object is, usually the more mass it has, and the more energy it has (or capacity to push back and attract). But not always, and that's the issue.

The reason Newton's gravity is solely an attracting force, while in electrostatic, Coulomb's Law is both an attracting and repelling force, is because they are measuring different phenomena, although both phenomena are linked to gravity. These two laws are not really interchangeable, despite sometimes scientists interchanging them in certain situations, while probably still getting relevant results.

I always thought Mark McCutcheon had been extremely specific when he developed his Atomic Expansion Equation to measure gravity. He did not say, gravity is the *reduction* in distance between expanding objects, he stated that gravity is the *changing* distance between expanding objects. As if to take into account this counter interaction of the orbits, or any attracting or repelling force, counteracting the attracting force of gravity, or even simply the momentum an object already has, while being affected by gravity. Once this begins to be applied in our science, and an army of physicists develop this further, we should see very interesting results.

In Newton there is a distance reduction due to the inner expansion of the objects. And in Coulomb, there is an additional attracting or repelling force countering this distance reduction, due to the orbits of *smaller* particles or objects, linking together but keeping apart, *larger* particles or objects. This additional attracting or repelling force counteracts the first distance reduction due to the inner expansion, through constantly enlarging orbits keeping the particles or objects apart.

For example, solar systems can attract or repel each other depending on the speed at which the rogue planets are orbiting between them. The faster the orbits, due to more matter being added to the system or pressure added, and the more spaced out the orbits will become, seeming to be a repelling force between solar systems.

The slower the orbits of the rogue planets, due to matter leaving the system or being destroyed, or the pressure being relieved, the smaller the orbits will become, seeming to be an attracting force. All of this while still counteracting the distance reduction due to their inner expansion, in other words, due to gravity. Solar systems are electrons, it is the same whether we are talking at the smaller scale or the larger one.

Coulomb's Law measures this force keeping the objects apart, while Newton measures the overall distance reduction between all objects due to their constant inner expansion, which never stops, even in a cloud of electrons completely immobile between two charged rods in static electricity.

There are no more charges or charged particles in Atomic Expansion Theory. Electrons can attract or repel each other, depending on the speed at which the neutrinos linking them together are orbiting. The faster the orbits of the neutrinos - due to more energy being added to the system, like more electrons through heat for example, or added pressure being added by squeezing the entire system - the larger the orbits of the neutrinos will be, seeming to be a repelling force. And vice versa for the slower orbits of the neutrinos creating an attracting force between the electrons.

The same goes for galaxies. The rogue solar systems orbiting between them and linking them together, can orbit faster in longer orbits, the galaxies are then being repelled, or orbit slower in shorter orbits, the galaxies are then attracting each other. And yet, all these galaxies and solar systems continue to expand constantly, and their orbits are constantly enlarging in an acceleration, to compensate for the accelerating reduction in distance due to gravity.

Galaxies are atoms, and galaxies linked together by solar systems are forming molecules at a larger scale. It is the same for atoms linked together by orbiting electrons, together they compose different molecules. Eventually we might use Coulomb's Law, or some similar equation, applied to the larger scale, to measure the force between solar systems linked together.

To summarise, Newton's attracting force, or gravity, will always reduce the distance between objects due to their constant inner expansion. However, gravity can be counteracted by particles orbiting between objects, keeping these objects apart. These orbits can accelerate and enlarge, or decelerate and shrink, changing the distance between these objects while keeping them apart, despite the considerable expansion of all these objects. And this is Coulomb's attracting or repulsing force. This is whether we are talking about electrons with orbiting neutrinos, or solar systems with orbiting rogue planets between them. Or atoms with orbiting electrons, or galaxies with orbiting rogue solar systems.

It is worth noting here that there are no forces involved per se. It is all an intricate system of smaller particles' orbits enlarging and contracting, being exchanged between larger particles, pushing back or pulling together depending on the circumstances. Newton's Law of Universal Gravitation has now been replaced by Mark McCutcheon's Atomic and Subatomic Expansion Equation. And the discussion above is how I re-interpret Coulomb's Law in light of Atomic Expansion Theory. I will discuss this in more detail later on in the book.

Another important point, is that an orbit does not really counteract the distance reduction of gravity. An orbit is just another outcome of the geometry of expansion, as much as the distance reduction in gravity is. However, in both cases, the distance reduces between everything, if not between the objects in orbits within a particular system, at least between the overall objects they form, and other such objects elsewhere in their surroundings. Like the distance between two groupings of linked solar systems, for example, or two clusters of electrons, which would not be linked together by orbiting electrons.

In *New Age Physics,* when applied to ultraterrestrials and their physics, you add to this the frequency or speed at which the neutrinos and electrons are orbiting everywhere, which affects the size of these particles and how fast they vibrate, establishing in which dimension these objects exist in.

There could be many planets, solar systems and galaxies that we don't even see or can interact with. It is doubtful they could have much impact on what we can see and interact with, but who knows. This is however not the solution to resolve the missing dark matter of the universe, since this dark matter was only invented for Einstein's theories to still work, which we no longer have any use for in Atomic Expansion Theory. We can all stop looking for dark matter or dark energy, the sooner the better.

2.13 A different concept of positively and negatively charged particles in Atomic Expansion Theory

Only the particles exchanging smaller particles orbiting between them, can be considered charged particles, positive or negative, as only they can either attract or repel other particles. The same goes for solar systems locked in with other solar systems (electrons) through orbiting planets between them, and galaxies (atoms) with orbiting solar systems between them. After that, a simple natural orbit effect between two particles, or the failure of one materialising for whatever reason, might also be interpreted as an attracting charge or neutrality.

There could also be an exchange of several neutrinos causing an attracting or repelling force, even when dealing with only two particles, such as an electron and a proton. An electron has orbiting neutrinos, and a proton is full of electrons with orbiting neutrinos, that can push back or attract simply by exchanging neutrinos, and by changing the speed and the size of the orbits of their neutrinos, whenever other particles come close.

As to why neutrons are neutral particles, and don't appear to be exchanging smaller particles with other particles they encounter, must have something to do with how many electrons and neutrinos they have. A configuration which neither pushes back nor attracts other particles, or where, perhaps, no exchange of particles take place.

At any rate, the very idea of protons and neutrons will need to be re-thought, because I don't see them when I look at a galaxy. And where are those quarks? I only see groupings of solar systems bonded together by rogue planets, each group representing probably a certain amount of energy, that we have identified as particles such as protons and neutrons. If two groupings don't exchange any planets between them, because of their configuration, number of planets, or very likely because of relatively large distances separating them, then possibly they could be considered neutral, neither attracting or repelling other particles.

The only true neutral particle in nature should be the neutrinos of all sizes, in my opinion. They are the only particles that do not exchange any smaller particles between them. But neutrinos do have orbiting smaller particles around them, just like planets have moons. It is just that planets don't exchange their moons. However, it could be possible, and maybe we will witness this one day, if we ever visit other solar systems, that planets might have moons orbiting between them. If it turns out to exist, even the neutrinos could not be considered neutral. Only moons and sub-neutrinos would then be considered neutral particles.

It is in view of this that I thought it would be better to make the distinction between neutrinos and sub-neutrinos in the very small. There are neutrinos, our planets, and there are even smaller particles orbiting them, the equivalent to our moons. I thought we should recognise them and give them a proper name, because they might become important in the future. And of course, I decided to call suns super-neutrinos. I think Seth, in Jane Roberts' books, did say that we would eventually discover a new particle even smaller than the neutrino. If it was not in Jane Roberts, then it was in *The Urantia Book*. And this is all becoming a self-fulfilling prophecy.

As in the discussion in the previous section shows, such particles are not really charged particles, they simply contain either more or less smaller particles in higher orbit than their neighbours orbiting around them. If a neighbour comes close enough, and it has less particles than its neighbours, some smaller particles will start orbiting between both. And the natural orbit effect of the geometry of expansion will ensure a complete balance of all systems, changing all the orbits instantly whenever new particles pass by or go away. The natural orbit effect of the geometry of expansion will be explained in more detail in chapter 3.

Wherever there is a surplus of the smaller particles, the orbits will re-arrange themselves to patch any other place where there is a depletion. The orbits will equalise in size between systems, either enlarging or reducing their orbits to balance. And the speed of the smaller particles will also equalise, either going faster or reducing in speed. Again, this is all due to the relative motion of the geometry of expansion.

This is the attracting and repulsing force that bonds the atoms together to form molecules, and which bonds together electrons to form larger particles such as protons and neutrons, clusters in light and heat, and clouds in magnetic fields. This is the reason some particles are said to be positive or negative.

If there is a same number of smaller particles orbiting the larger particles, they will repel, as there is nothing to balance. And once balance has been achieved, these particles or systems are locked in through these orbiting smaller particles. This is when we have stable particles or molecules.

A charged particle, one that is either positive or negative, capable of attracting or repelling another particle, is no longer really a charged entity possessing an electric charge of some sort. It is just a particle or a system with more or less smaller particles orbiting it, and orbiting between the particles and systems, with an elaborate natural orbit effect to balance the overall system. An electric discharge is all the smaller particles becoming unstable and going everywhere, or more specifically wherever there are sudden large depletions of these smaller particles, like in lightning.

Neutrinos, which are planets, and sub-neutrinos, which are moons, don't exchange smaller particles or objects between them, as far as we know. This lack of exchange of particles between them, means that they cannot attract or repel anything. They are considered neutral particles and objects, although they are still subject to the natural orbit effect of the geometry of expansion. And in their expansion, the distance between them will reduce, unless counteracted by their enlarging orbits.

In solar systems where there is more than one sun, or in electrons where there is more than one super-neutrino, suns and super-neutrinos are exchanging particles between them, so they are not neutral particles, unless there is only one of them in their system.

An electron has orbiting neutrinos within it, and can attract other electrons passing by through an exchange of neutrinos, if there is an imbalance of neutrinos between them. The same for solar systems who attract or repel other solar systems passing by, depending on the number of planets overall orbiting within the solar system, and between other solar systems already linked to it.

An atom has orbiting electrons, and is exchanging electrons with other atoms nearby. The number of orbiting electrons the atom contains, and that are being exchanged already with nearby atoms, will determine, through the natural orbit effect of expansion, if any new atoms passing by will link to the system or not, or if it will be attracted or repelled. This is what chemical bonds are and what defines the Periodic Table of Elements, and what kind of element the atoms will end up being. The same goes for galaxies with their orbiting solar systems within the galaxies and between galaxies. They will either attract or repel other galaxies and form cosmic molecules and elements at a higher scale universe.

Would you not like to know what element of the Periodic Table of Elements our galaxy is? And the molecule that it forms with its neighbours at a higher scale? And what about our solar system, is it linked to other solar systems around, in order to form the energetic equivalent of a proton, or a neutron? Although they look nothing like the balls or point particles we imagined them to be. With this new information, someone out there eventually might find out and tell us.

One last important thing. There is this idea floating out there that if the Sun were to suddenly be destroyed, this information would need to travel in space to the very last planet of the solar system, before that last planet could start changing its orbit. And there were discussions about gravitational waves or graviton particles, capable of travelling faster than light, which would transmit this information instantly. However, none if this is required.

The geometry of expansion is instant everywhere. As soon as the Sun would be destroyed, instantly all the orbits would change, not only in our solar system, but across many other solar systems linked to ours. It could even affect the orbits of everything in the galaxy, and perhaps beyond. It would be instant, because this relative motion of the geometry of expansion is all due to what can be seen behind the scenes, where we can see the expansion.

Behind the scenes, all we see are expanding balls moving in all directions, no orbits whatsoever, and it is the fact that we cannot see this expansion that results in our reality of orbits. If one dot disappears behind the scenes, it instantly changes everywhere the entire geometry of our resulting reality. No other dots need to know that one has disappeared, none of these other dots changed their path or speed as a result of one dot disappearing. They just continue on their existing paths and their expansion, which instantly changes everywhere the geometry of the resulting universe we see, moment by moment.

2.14 Interdimensional time is relative - Time is not relative nor a dimension as Einstein claimed

Time is no longer relative, and certainly no longer a dimension as in Einstein's concept of spacetime. Time simply does not exist, except as a human convention, for practical calculations that make sense to us. Time essentially runs the same everywhere, within our dimension. For the entire debunk of the proofs of Einstein's concepts, please read the free excerpts mentioned at *1.9 Debunk of Standard Theory.*

As soon as you change dimension however, indeed the time rate appears to change. I must stress that what truly changes here, is simply the perception of the passing of time. Most importantly, your watch and atomic clocks are not going to run faster or slower no matter what. They will always tick at the same rate, unlike in Einstein, despite the numerous proofs to the contrary.

Someone vibrating faster in a higher dimension, will be able to achieve much more than us within a same amount of time. Days could go by for them, within a few minutes here, but it is only a different perception of time. For example, the Earth is not going to start orbiting faster around the Sun just because you moved to a higher dimension, and suddenly can do things much faster than normal, or accomplish more within a same amount of time.

Have you ever wondered how a colleague at work, seems capable of doing twice more work than you in a same amount of time, while you thought you were actually quite quick, and could not possibly work any faster? How is this colleague doing it then? Is he or she vibrating at the same rate as you? I had several of these colleagues in the past, who cleared so much work in record time, they appeared to defy the laws of physics.

It is like a fly vibrating at 250 hertz compared to us vibrating at 50-60 hertz. They can do more than us, up to four times more, in a same amount of time. For us they move in fast forward, just like for us tortoises move in slow motion, as they vibrate at a frequency of 15 hertz.

UFOs are not the only things flying around in weird paths, and often at a right angle, stopping in midstream. Just look at the flight patterns of flies, and you will observe that they often move in the exact same way, often at a right angle and stopping in midstream. This could be only how we perceive it, because essentially flies are living in a different dimension than ours, one that we can still see and interact with, where everything is vibrating faster, and where time is perceived as ticking faster. However, what we see might not be what really is, due to motion being relative.

From witness accounts, aliens, or ultraterrestrials, have eyes that look more like lenses, and they are reminiscent of the eyes of flies and other insects. There is an excellent reason, as to why they have a similar kind of eyes. Vibrating faster, they see light of a higher frequency, but from a higher dimension. This is why the most efficient fly zappers use ultraviolet light, in order to attract the insects into their trap. What they see, must be similar to our ultraviolet light, without being exactly the same. The clusters of electrons within their light, must be larger, more spaced out, with neutrinos and electrons orbiting faster. That kind of light lays outside our range of visible light.

Lizard eyes, like the ones of demons or other creatures from the lower dimensions, must be more sensitive to infrared light, because reptiles live in slow motion and vibrate slower than us. If our matter were to vibrate a little bit faster, we would never see such creatures, they would disappear into their lower dimensions. Plus, we might meet more beings vibrating at a higher frequency of matter.

This also explains those flying rods, that we can only see in slow motion on films or photos. They are most likely insects vibrating so fast, we can't see them with the naked eye. We need to first film them and view it in slow motion. And even then, we only see a spiralling elongated shadow of what they truly are.

You can see this "rod" effect with the naked eye just by looking at normal insects outside in the summer. Some are buzzing so fast in swarms, you can only see elongated blur images that look precisely like images of flying rods. But once they stop, you can see what they truly look like, normal insects. I even personally observed insects disappearing completely from view over short distances, until they stopped their flight, it is not uncommon.

And now you know why insects and flies can easily defy gravity and remain and walk on the ceiling. They weight almost nothing, the matter composing them is less dense, and the ceiling expanding downward, with everything else that is expanding in the universe, is enough to create a small gravity effect that can keep them firmly on the ceiling. In its downward expansion, the ceiling pushes down on the insects, keeping them on the ceiling. It also works with larger bugs, when sometimes it seems they are defying the laws of gravity, because they are of a lesser density while vibrating faster.

Anything of a lesser density is pushed upward by what is denser. If bugs are lighter than air, or of the right density to feel almost weightless, then they can easily stick to walls and ceilings. In some cases, they can even levitate and fly long distances without the need for wings, as in the mystery of the flying spiders covering miles over the sea to reach boats. And also, the flying Tibetan Yogis capable of mentally altering their vibrational rate, in order to become less dense than air. This is the kind of thing our technology needs to exploit, so we can fly out into space without any need for propulsion.

On a more personal level, in order to vibrate faster, without becoming obsessive, you only need to think of it and imagine it happening. Feel lighter and think more positively about life, and the world we live in, despite all the fears and how bleak it might look. I'm no expert, but feeling love for everything, or feeling something more positive, should help. This is what inspiration, motivation and creativity are all about, it fills you with energy. Freedom and liberty are a trick of the mind, you are only ever as free as you think you are. If you feel free, you vibrate faster, and all the negative does not affect you emotionally.

2.15 $E=mc^2$ is just the equation for the kinetic energy of a moving object, known prior to Einstein

Einstein's famous equation $E=mc^2$, is just the known equation for the kinetic energy of moving objects in disguise, as Mark McCutcheon discovered while doing its derivation. The classical equation for the kinetic energy of a moving object, that existed prior to Einstein, is $E_k = ½mv^2$, which states that the object's kinetic energy, E_k, is equal to one-half of its mass times its velocity squared.

The difference between the two equations, concerns how the force of impact of light can double, when hitting and bouncing back reflecting surfaces such as mirrors. This said, kinetic energy has been described as the only real type of energy by Mark McCutcheon, but it is the force of impact that particles and objects can acquire through relative motion, and it only involves matter, like the clusters of electrons composing light. Light is not made of energy in Atomic Expansion Theory, light is made of matter.

From this very short discussion, we can immediately see what Einstein's equation is. First, you need to remember that mass is made of molecules, which are made of atoms, and atoms are made solely of electrons. Then, light and heat are made of clusters of electrons. E is the kinetic energy, or force of impact, of a mass m, instantly turned into clusters of electrons, or light, multiplied by the speed of light c. c is simply the high speed at which clusters of electrons are launched out of the atoms, when they are broken down, plus any acceleration due to gravity before they hit something. And the square means an acceleration, in other words, that the effect is compounding.

The acceleration comes from gravity, the dual decrease in distance between these clusters of electrons exploding in every direction through a chain reaction, and everything else. It is their speed or motion, which defines their force of impact when they hit anything, which is the definition of kinetic energy.

There is no ethereal concept of energy here to speak of, it is just the force of impact of moving matter. The kinetic equation for light did not require an upgrade, as it was perfectly fine as it was. And it highlights a significant problem with $E=mc^2$, the fact that the force of impact of a nuclear bomb, can double when hitting for example mirrored buildings. His equation should instead be a range between $E=mc^2$ and $E=2mc^2$, to take into account the reflecting surfaces of each object that might be impacted by radiation.

2.16 Redshifts misinterpretation - No dark matter, dark energy, Big Bang or black holes

New Age Physics is a new science that could revolutionise everything, overall it is called Atomic Expansion Theory. I would call it Expansion Theory, but people would think I am referring to the

Theory of Inflation of the universe. They think the universe is ever expanding faster, because of dark matter and dark energy, which are inventions in a botched attempt to make Einstein's equations still work, when it is clear to all that his theories are simply incorrect, by a long shot.

The redshift in light is nothing like sound, there is no Doppler effect in light. Redshifts are no indication of distance of astronomical objects, nor of speed, nor of direction of movement. Things are not moving ever faster away from us, quite the contrary, the observations show that the universe is more static than anything else.

There is no dark matter, there is no dark energy, and there was no one Big Bang that started our universe, along with its singularity where space and time lose their meaning. After all, no one has ever been able to identify a central point in the universe, from where everything would have originated from. There is no need for them once you understand how we misinterpreted redshifts.

There are no such things as black holes or wormholes either. A black hole is simply a dead star that went supernova some time ago. These dead stars can be extremely large and have a significant effect on the orbits of everything around them, while their proportional significant expansion will catch up with a lot of matter near them. However, they are only black from our point of view, because dead stars don't emit light. It makes no difference whether the star is alive and shining, or dead without emitting any light, the gravitational effect is the same. There are no black holes at the centre of galaxies either. Anyone who looks carefully at any galaxy, will notice that the middle appears to be more like a massive sun than a black hole. And if not, then you are looking at a dead massive sun.

2.17 The double slit experiment misinterpretation and the issue with particle accelerators

The double slit experiment, which is at the base of the entire Quantum Mechanics, has been misinterpreted. It is said that one electron being ejected in front of two little openings, will sometimes interfere with itself, and create a wave pattern on the screen behind the openings, instead of dots as it should be, if they were to pass through only one slit. This is said to be a proof that the electron is going through both openings at the same time, proving the wave-particle duality of matter. Moreover, it is said that whether we observe the experiment, or don't, will alter the results.

For Mark McCutcheon, the author of *The Final Theory* and Atomic Expansion Theory, it is simple. When they say that only one electron is being ejected at a time, before passing through one slit or the other, or through both slits at the same time, he says that in fact there are more than one electron being ejected, but our measuring instruments cannot detect them. Hence, they still have the interfering pattern of waves on the screen behind the slits. And of course, maybe a whole bunch of neutrinos are also ejected with each electron.

According to Mark McCutcheon, it would prove almost unthinkable that we would be able to fire off just one electron or one cluster of electrons (photon) at a time. In his debunk of Quantum Mechanics, McCutcheon shows that energy or light does not have a dual nature of wave and particles. Instead, it is always a sea of particles that move in waves. Just like molecules of gases will move in waves of air, and water molecules will move in waves of water. Waves of energy cannot cancel each other out, it would violate the Law of Conservation of Energy. Instead, simply all the particles stop moving in waves and scatter in all directions.

As to the results being different depending on if we are observing the experiment or not, if the act of observing requires any kind of particles being thrown at the experiment, such as light, which are clusters of electrons, or beams of electrons, then of course it would affect the electrons or clusters being fired within the experiment. Not observing the experiment will result in less particles going around in there, observing it might flood the whole experiment with particles, altering the results.

This is also a problem in particle accelerators. If you are using electricity or electromagnetic fields to accelerate the particles in an accelerator, then you are essentially firing clouds of electrons to speed up the particles. Their expansion rate is set, the speed at which they are launched freely into space will never push them ever faster. It does not mean that the mass of the particles is becoming infinite, as Einstein's relativity states, and that they can no longer go over the speed of light. The experiment is simply limited by the speed of the particles they use in order to accelerate them.

And of course, if all particles are made of electrons, and you are constantly throwing electrons at them, bits will break up, others will stick and create what seems like new particles that will only exist for a micro-second. However, these are not new particles, you can easily create an infinite number of particles if you wish, all of them with a slightly different number of electrons. Only stable particles in nature are worthwhile and should be considered particles.

You will note that beyond neutrinos, electrons, clusters of electrons, protons, neutrons and atoms, neither Mark McCutcheon nor I mention any other particle. McCutcheon does not even mention the neutrinos. And even then, for me the protons and neutrons at the larger scale, are now more like shapeless groupings of solar systems, linked together by orbiting planets between them, which overall correspond to a certain energetic value equivalent to a proton or a neutron at the smaller scale.

This must reflect what protons and neutrons look like in the very small. Maybe they have a shape, maybe they are balls. We would need to map accurately a few solar systems in our galaxy, not using redshifts in light, and check which ones are linked together through orbiting planets, and identify what shape these groupings might have.

2.18 Quantum Entanglement misinterpretation

Quantum Entanglement is a strange phenomenon where two subatomic particles from a same source, separated in space, are interconnected somehow. And if you change the polarity of one particle, the other change accordingly instantly across a huge distance, and faster than the speed of light. How can that be?

As Mark McCutcheon says, with the new understanding, the nature of light is radically changed from separate photons fired through space, to continuous beams of expanding subatomic-matter clusters that our eyes detect, to generate the experience of colour and brightness.

In this case, this is not an experiment with two photons exhibiting mysterious "quantum entanglement", but merely two separate unseen continuous beams of expanding matter clusters, physically connected back to where they were split, from one initial beam. Then, the more likely explanation of the "entanglement" effect, is that an influence altering one beam, is conducted along this continuous span of unseen physically connected matter clusters, which goes on to affect the other beam.

Since vibrations in solid objects travel faster the denser the material, the speed of conduction through the extremely dense span of such subatomic-matter clusters in light, may well be extremely rapid - even far exceeding the speed of light. The "entanglement" experiments appear to suggest this possibility of conducting signals along beams of light at speeds that so far appear to be instantaneous, providing a practical possibility for faster-than-light communication. And this certainly could speed up computer processors, once we have the correct understanding of what is really going on.

2.19 Parallel universes - The Heisenberg uncertainty principle misinterpretation

The Heisenberg uncertainty principle is the assumption that we can only tell either the position of an object, or its speed or momentum, but not both in any accurate way. Atomic Expansion Theory explains everything Quantum Mechanics explains, and demystifies the misunderstandings that led to the theory being called spooky by Einstein himself.

For me, we can tell the position of an electron accurately and its speed, both at the same time. Because an electron is a solar system, and when we look at solar systems in the universe, they are not moving very fast from our point of view. We can tell the speed they are going at, and their position. It is just that when things are moving so fast at the smaller scale, from our perspective, it becomes extremely difficult to make measurements.

The main issue with the uncertainty principle, is that we are using light or beams of electrons in order to observe and measure something itself made of electrons. Light is made of clusters of electrons, which probably comes with a lot of loose electrons, which destroys the delicate thing we are trying to observe at such a small scale. To observe anything in the very small, we would require measuring instruments which are not throwing away particles of any kind towards what is observed, otherwise of course we will affect the entire results of what we are observing.

About superposition, we might very well think a particle is at more than one place at any given time, simply because of the limitation of our measuring instruments. It is just going too fast. They say it is not related to the limitation of our instruments, because of Erwin Schrödinger's equations, but I think it is. Because quite possibly these particles are moving faster than the speed of light, perhaps multiple times from our perspective, and this is why we might see the same particle at more than one place at the same time, when in fact it is only in one place at any given time. I believe it is even possible, that without going at the speed of light, something moving extremely fast might appear to us to be in multiple locations at the same time. So much for the possibility of a multiverse, then.

This is a very simple description of a much more complicated topic, which involves Einstein's relativity as well. If you read Mark McCutcheon's debunk of Quantum Mechanics in his book *The Final Theory,* it becomes clear Quantum Mechanics is an incorrect theory. Especially when Atomic Expansion Theory can explain it all without spooking anyone. For a start, as mentioned above, McCutcheon proved that matter, energy and light don't have a dual nature of wave and particles, they are all particles that sometimes move in waves.

Reading the file on my website of Mark McCutcheon's debunk of Quantum Mechanics, is not going to be enough to prove what I just said, because the file omits the Atomic Expansion Theory's explanations of the descriptions of Quantum Mechanics. You will have to read his book for a more complete picture.

This said, there can still be parallel universes. If this reality is psychological or psychophysical in nature, and we are all creating our own bubble universe like computers on a network, and that everything is built pretty much like collective dreams, then you are still free to think and create any reality you wish to live in. You are only limited by your doubts and what you think is possible, and you can create as many parallel universes as you can think.

Seth in Jane Roberts' books, does state that reality constantly splits into several others, which are then expanding in all directions. I imagine it would be like several instances of the same computer simulation, all slightly different, all running at the same time, that our will can influence, since we are like little biological computers all interconnected on a collective network. So, hacking reality appears to be possible, Jesus-Christ must have been a hell of a hacker.

2.20 Five types of expansion and contraction of matter to explain the entire physics

New Age Physics is a complete theory of everything based on five different types of expansion and contraction of matter, which should open the doors of science to rival anything related to ultraterrestrial technology, including levitating large stones and spaceships through the buoyancy of matter of a lesser density, traveling anywhere in the universe instantaneously, and even time travel.

The same laws of physics can now be applied to both the very small and the very large, where moons, planets and suns are neutrinos, electrons are solar systems, atoms are galaxies, and clusters of galaxies form cosmic molecules, composing objects and beings at a yet higher scale universe. Just like in particle physics, we will find that rogue planets orbit between solar systems, and rogue solar systems orbit between galaxies and clusters of galaxies, linking them together to form larger objects.

At the smaller scale, although there must be an even smaller scale universe, the neutrino is the only fundamental particle in existence, all other particles are made of neutrinos. The changing speed at which neutrinos and electrons are orbiting, within and between larger particles, and the varying size of their orbits as a result of their speed, correspond to the interdimensional frequency or vibrational rate of matter, defining the interdimensional density of matter.

The five different types of expansion and contraction of matter explain the entire physics governing all realms of existence, even the realm of dreams and thoughts:

- The first type is the constant inner expansion of matter – the neutrinos, electrons, atoms, molecules, planets, solar systems, galaxies and the universe – expand at a constant universal rate of 0.00000077 of proportional size per second square, causing the Earth to expand by 4.9 metres each second. This explains gravity and orbits through the geometry of expansion, gravity being the changing distance between expanding objects. This inner expansion remains unseen to us, since we also expand. Likewise, it elucidates the strong nuclear force – which is no longer required to keep the nucleus together – as simply protons and neutrons in the atom's nucleus, pushing against each other in their expansion. While the weak interaction, or radiation, is due to a neutron being an unstable particle by nature, which decays into a more stable proton, an electron and neutrinos.

- The second type is the growth and shrinkage of the electrons, when the neutrinos orbiting within and between the electrons, get excited or settle down, causing their orbits to enlarge or reduce in size. This is the crossover effect, when electrons *move out* of the atom to push and pull against each other in their inner expansion, equalising in size when in contact. Through the geometry of expansion, they move from where there are surpluses of electrons to where there are depletions. Moving either in clouds, clusters, bands or lines of electrons, it goes on to explain electromagnetism, attracting and repelling forces, heat and light, radio signals and electricity. The first two types of expansion and contraction of matter, unite all four main forces of nature into a theory of everything.
- The third type is the growth and shrinkage of the electrons *within* the atom, leading to physical realities with larger or smaller atoms and molecules, more or less spaced out, and of different vibrational rates and densities of matter, explaining other dimensional worlds and beings.
- The fourth type is the instant expansion of our psychophysical individual bubble realities, from our

consciousness and brain, like computers do on a network, to build an interconnected collective reality, more akin to a virtual world. We cannot see or interact with other dimensional worlds, because we are not attuned to these frequencies of matter, just like a radio picks up only one station at a time. Our brain and consciousness do not create these worlds for us to see within our bubble reality, our particles would first need to vibrate at these specific frequencies.

- And the fifth type is Universal Relativity, the expansion and contraction of matter due to relative motion. Objects only ever expand or reduce in size as they move, in a universe without distance, where everything is located here around us. This would be expected in a psychological or virtual reality, or a simulated world or holographic universe. We can move anywhere instantly at any particular time in history, by simply vibrating at the frequency or resonance of that particular time and place. In other words, by adjusting the speed of our particles, which defines our relative size, our location, the date and the interdimensional time rate. This is how we will instantaneously travel in the future.

This is just the core of the theory, the beginning of what it all means, and how it could change the world. This is what *New Age Physics* is about, the exploration of those five types of expansion and contraction of matter, which explain our entire physics in an orderly theory of everything. We no longer need a PhD in science to understand the physics of our universe.

THREE

The Final Theory by Mark McCutcheon

3.1 First and second types of expansion and contraction of matter according to Mark McCutcheon, concerning General Relativity and the four fundamental forces

This third chapter is a very condensed version of Mark McCutcheon's book *The Final Theory.* I could not here explain why or how everything he says is justified, I felt there was no need to re-write his book, where everything is so well explained already. For the proof, there is no substitute to actually reading his book. You can even read it on Kindle or Apple Books, and it is cleverly written for people with no prior knowledge of our actual physics.

What I came to call the first and second types of expansion and contraction of matter, that follow from Mark McCutcheon's book, would be true even if everything else I developed from his theories was incorrect. On its own, Mark McCutcheon's Atomic Expansion Theory, is the long sought-after theory of everything that we have been searching for. I have made it clear whenever we diverge in our theories by stating it outright, or with indications such as *for me, I believe, I think* and *in my opinion.*

3.2 Atomic Expansion Theory

Atomic Expansion Theory is a working unified field theory uniting the main four interactions of nature: gravity, electromagnetism, and the strong and weak nuclear forces of the atom. This theory of everything has one principle only, the constant expansion of the atom, which is the first type of expansion of matter. This new principle also states that there is only one fundamental particle in existence, it is the electron. Everything is made of electrons constantly expanding at a certain rate, which is also the first type of expansion.

Atoms, any particle, energy fields, magnetic fields, radio and TV signals, heat and light, are all and solely made of growing and shrinking electrons, which is the second type of expansion and contraction of matter. And from there we can explain the entire physics, while uniting both the physics of the very small and of the very large.

3.3 New model of the atom, strong and weak nuclear forces and chemical bonds

According to Mark McCutcheon, the electrons bounce off the nucleus of the atom instead of orbiting it. However, I personally believe they are still orbiting, and I will explain this further later on. The protons and neutrons within the atom are made of expanding electrons, and they push against each other in their expansion. The nucleus also constantly expands, explaining why the bouncing electrons are reaching higher every time by half an electron, creating that constant small expansion of the atom at the edge. The atoms coming closer to each other, bond together by exchanging bouncing electrons between their nuclei, and this is how molecules are formed, the chemical bonds.

This elucidates the strong nuclear force within the atom. There is no need for an invented nuclear force within the atom, in order to keep the protons and neutrons together, and prevent the nucleus from flying apart. It is easily understood, within Atomic Expansion Theory, that these particles are simply pushing against each other in their expansion.

A neutron is a less stable proton, and a neutron becomes a proton by simply losing one of its electrons, while the electrons composing it are trying to find a balance and become more stable. This describes the weak nuclear force, but once again no invented force was required to explain this phenomenon. All this is very well explained in Mark McCutcheon's book, where all the equations are derived in optional maths sections. All four main forces of nature can be explained in a logical manner through the expansion of matter, the other two being gravity and electromagnetism.

3.4 Gravity and the formation of galaxies

The bouncing and expanding electrons within the atom, are responsible for the small expansion of the atom, and this expansion is gravity. As objects expand in the universe, the distance between them reduces. All objects in the universe, no matter the scale, expand at the same constant rate (X_A) of 0.00000077 of proportional size per second square, or 7.7×10^{-7} /s^2. As per Mark's mathematical equations, everything in the universe doubles in size every 19 minutes.

This might seem alarming and unconceivable, despite the incredible vastness of the universe. However, this expansion happens behind the scenes, meaning that it remains unseen by us, because we are expanding as well. As a consequence, expansion is not part of our resulting physical reality. All we see are shrinking distances between objects which appear to remain the same size.

The expansion of Earth underneath our feet, behind the scenes, is what keeps us on the ground. Distance shrinks between objects because objects expand, not because there is a mysterious attracting force acting at a distance, which would be emanating from matter, as in Newton's Law of Universal Gravitation. Or because objects distort the fabric of spacetime like a rubber sheet, making smaller passing objects to follow the curves made by larger ones on that sheet, as in Einstein's Theory of General Relativity.

I believe the galaxies and the universe also double in size every 19 minutes, but Mark McCutcheon does not believe so. From the way he describes the formation and the workings of galaxies, he feels galaxies are not expanding. As a consequence, neither is the universe.

He says, as everything expands within a galaxy, distances shrink and the galaxy reduces in size. This causes all the solar systems to move towards the centre, like water draining down a hole in a sink, explaining their spiralling arms formation as they move down the centre. All the while, new solar systems are always being created at the edge of the galaxies, so galaxies remain more or less the same size. It has been confirmed that new solar systems appear at the edge of galaxies. And the clusters of galaxies, as can be observed in the night sky, are more or less static, they are not moving away from us at an alarming rate.

However, I have another way to explain the spiralling arms formation within a galaxy, which does not involve the analogy of draining water down a hole. It just shows how easily we can all find different ways to explain the same phenomena in physics, justifying the several different models we now have to explain the universe we live in.

The spiralling arms formation of solar systems, are not due to them moving towards the centre as all distances shrink, like draining water in a sink. Instead, I feel solar systems and their arms are simply orbiting around the centre of the galaxy, and if anything, behind the scenes, solar systems are actually moving away from the centre in enlarging orbits.

The spiralling arms formation are simply due to an initial explosion at the centre of the galaxy, where too much pressure built up from too large an object. And the only way all the matter could escape that pressure, was by squeezing out at both poles of the massive object, creating two long straight arms of matter being ejected out on both sides. And the process can repeat again later on, if too much pressure builds up again, then two new straight arms will appear. In time, through the inner expansion of matter, solar systems will form within these arms.

As all the matter starts to orbit around the nucleus, the matter nearer the centre orbits faster, while all the matter at the edge orbits slower. Just like the planets around the Sun are orbiting faster the closer they are to the Sun, or a satellite in a low orbit around the Earth goes faster than one orbiting at a higher orbit. And since the solar systems' orbits are not very fast, as it takes between 225 and 250 million years, for solar systems to revolve once around the galaxy's centre, we don't see circular or elliptical orbits. What we see, are these solar systems shaped in spiralling arms instead, which are constantly twisting and stretching more and more as time goes by, because the matter closer to the centre is moving faster.

However, all solar systems must be orbiting in circular or elliptical orbits around the galaxy's centre, and not draining towards the centre following the spiralling arms. And these orbits, behind the scenes, are enlarging away from the centre at a constant rate. Moreover, the solar systems are interconnected together by an exchange of planets orbiting between them, so they remain more or less in their spiralling arms formation in their orbit around the galaxy centre.

In effect, galaxies also double in size every 19 minutes, and so is the universe. And clusters of galaxies would not need to move away from us at speeds faster than the speed of light for the universe to expand, since we cannot see this expansion. For us, all sizes, velocities and relative distances between everything, remain more or less proportionally the same, despite the entire universe doubling in size every 19 minutes behind the scenes.

This is important, because my findings reveal that galaxies are atoms while solar systems are electrons, and as such, they must also expand at the same constant Atomic Expansion Rate as everything else in the universe. Just as what we see in the universe are just cosmic molecules made of galaxies linked together by solar systems, composing something larger at a higher scale universe, where everything must also be expanding at the same constant rate.

If we consider the expanding-electron concept, which in turn leads to equally expanding atoms, a new gravitational theory emerges that actually mirrors Einstein's famous elevator-in-space thought experiment, where standing on Earth is entirely equivalent to being accelerated upward in space.

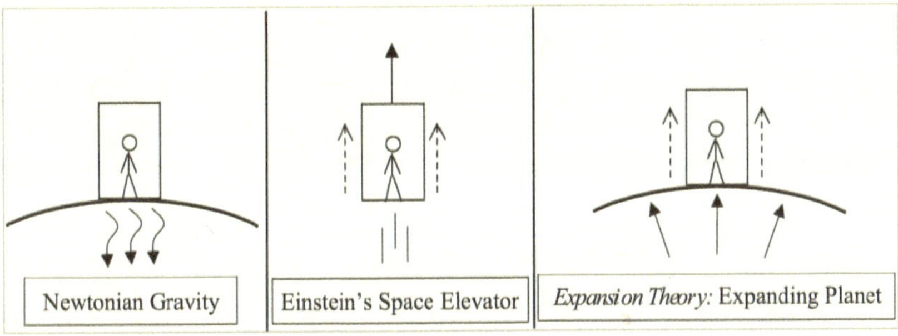

Figure 1 (2-3) Progression of ideas leading to Atomic Expansion Theory

In figure 1, showing the progression of ideas leading to Expansion Theory, we first see a man standing in a lift on Earth, being kept to the ground by an attracting force emanating from the Earth. In the middle image, we see Einstein's principle of equivalence, where being accelerated upward in a space elevator, is equivalent to Newtonian gravity. And then, in Expansion Theory, we see the man standing in the lift being pushed upward by an expanding planet.

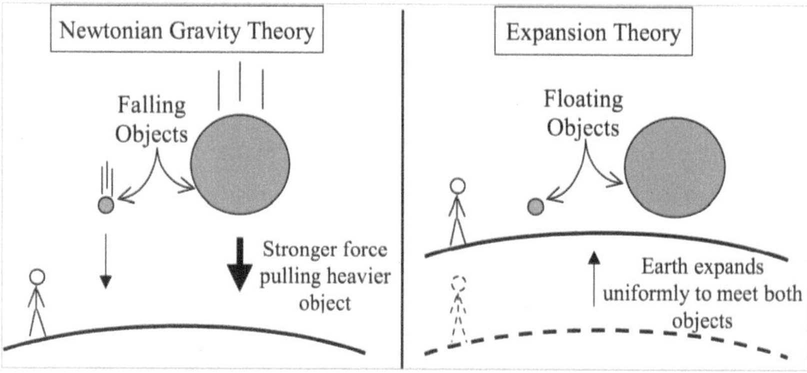

Figure 2 (2-2) Equally falling objects: Newton vs. Atomic Expansion Theory

In figure 2, we see that the force we feel underfoot is then due to our resulting expanding planet, with dropped objects all equally approached by the ground rather than the other way around, while the underlying expansion is unseen, as everything expands equally, maintaining a constant relative size. This would create the appearance of a force somehow holding us to the ground and pulling all objects equally downward, regardless of mass, just as Newton proposed, with a stronger force pulling heavier falling objects. And while Einstein opted for "warped spacetime", Atomic Expansion Theory suggests this far simpler and more literal possibility of objects simply floating in the air instead of falling, experiencing no g force, until an expanding planet catches up with them.

It is important to note that objects don't fall equally. Why a small marble and a cruise liner dropped from a certain height in a vacuum, will both reach the ground at the same time, is because from these insignificant heights, it is nearly impossible to measure this accurately.

According to Mark McCutcheon's equation for gravity, Atomic Expansion Theory would eventually show a difference for falling objects of different sizes. A larger object would reach the ground earlier, because it is proportionally expanding more than a smaller one, reducing the distance between the larger object and the ground faster. Just like in Newton, where a more massive falling object would experience a stronger force attracting it to the Earth.

Even in Einstein, a more massive object should reach the ground faster than a lighter one, because mass is part of his equations, as he described gravity as a result of space and time being curved by mass and energy. How could a massive object not be more affected by gravity than a lighter one, when more massive objects cause more pronounced curvatures of space?

In all three theories, I believe a bowling ball would reach the ground before a feather in a vacuum, no matter how many experiments proving otherwise you will find on the Internet. These experiments are too limited to tell. We would need to stop a planet in its orbit, place it immobile in space at a certain distance from the Sun, place a marble at the same distance on the other side of the Sun, and see if they both reach the Sun at the same time, or which one reaches the Sun first.

Where Einstein and McCutcheon differ from Newton, is that there is no force acting at a distance on any falling object, instead they are free floating in space. Both Einstein's Theory of General Relativity and Atomic Expansion Theory, explain gravity through geometry instead of forces. In Atomic Expansion Theory it is the geometry of expansion, while in Einstein objects fall to the ground following the curvature of space. But let's not forget that the falling objects also create their own curvature in space, which depends on their mass and speed (or energy).

Consequently, an overall larger curvature of space from both a planet and the Sun, should bring objects faster together than the miniature curvature a marble would create on the rubber sheet of spacetime. At any rate, Einstein stated his Principle of Equivalence between his theory of gravity compared with Newton's, and two massive objects will attract each other faster than a large and a small one, because in Newton the force of attraction is greater the more massive an object is.

The expansion of Earth is constant at 4.9 metres per second. The distance reduction between a free floating expanding object and Earth, is an acceleration of 9.8 m/s^2. People easily forget, when first encountering Atomic Expansion Theory, that the expansion of all objects is constant, and that it is only the distance decrease between expanding objects that is accelerating, as will be explained in the next section.

Hence, the higher you are in the sky, the faster and the harder the Earth will hit you. The air resistance you feel all around you is the air being pushed up in the Earth's expansion. This air resistance will eventually stop your acceleration when you reach terminal velocity. The atmosphere does not escape into space precisely because it is pushed by the Earth's expansion, creating air pressure and weather patterns.

3.5 Two types of distance decrease to calculate gravity

In Atomic Expansion Theory, Mark McCutcheon explains that atomic objects, like asteroids for example, are moving relative to each other. To calculate the decrease in distance between two objects moving in space, meaning to calculate gravity, his equation considers two types of distance decreases.

The absolute decrease in distance due to the own internal expansion of the objects, and also the further relative decrease in distance due to the fact that space between the objects does not expand along with the objects, since space does not exist as a concrete thing made of matter, and there is no ether either. Each moment that the objects have expanded further at their constant rate, means that there is an additional overall relative decrease in distance between the objects, from where the acceleration in gravity comes from.

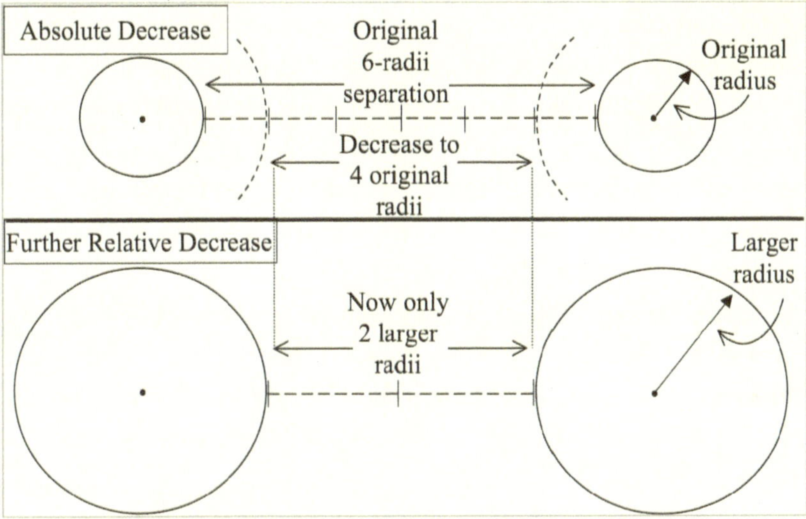

Figure 3 (2-6) Absolute and relative distance decreases

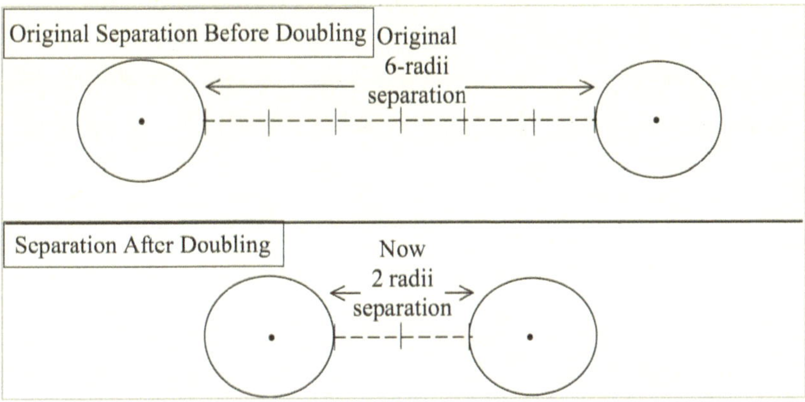

Figure 4 (2-7) Same scenario as in Figure 3 as it actually appears

Figure 3 shows the expansion behind the scenes when two identical objects double in size. In the top image, we see the original radius of the objects, and that six radii separate them. After they double in size, they are now separated by only four radii. This is the absolute decrease due to the constant expansion. However, when suddenly the two objects are twice larger, their radius is now twice longer, and consequently they are now separated by only two radii instead of four. This is the further relative decrease in gravity.

Figure 4 shows what happens as a result of it in our reality. Once again, above we see the two objects separated by six radii, and below, after the doubling behind the scenes, they are now separated by only two radii instead of four, just like at figure 3. However, because we cannot see the expansion, for us the radius is the same length as before. While behind the scenes, if we could see the expansion, this radius has now doubled in size. This explains the further relative decrease which needs to be taken into consideration, which is responsible for the acceleration of the distance reduction in gravity.

Objects in our reality can never be at rest, since in their constant expansion, distance is constantly decreasing between them, creating the illusion of a force of attraction. It makes no difference if the objects behind the scenes are moving, or are completely immobile in relation to each other. The expansion alone, even though there might not be any motion at all behind the scenes, leads to real motion in our resulting reality.

3.6 Atomic Expansion Equation to calculate gravity in Expansion Theory

Gravity, or the changing distance between expanding objects, is now explained by the simple inner expansion of all atoms and atomic objects. The dual distance decrease, mentioned in the previous section, is calculated using Mark McCutcheon's Atomic Expansion Equation. You will find the derivation, and how he found the universal atomic expansion rate X_A, in his book *The Final Theory,* which I am quoting from below:

3.6a The Atomic Expansion Equation

$$D' = D - n^2 X_A x (R_1 + R_2) / (1 + n^2 X_A)$$

Where:

D' is the changing distance between two expanding objects of radius **R_1** and **R_2**
D is the original distance between the two objects
n is the number of seconds that have passed since the original distance was measured between the two objects
X_A is the universal atomic expansion rate's constant 0.00000077 /s^2 or 7.7 x 10^{-7} /s^2

"The *Atomic Expansion Equation* above calculates the changing distance, ***D'***, between two expanding objects of radius R_1 and R_2 over time. The top portion of the equation is the *absolute* decrease in the original distance, ***D***, between the two expanding objects as they take up more space, and the bottom portion is the further *relative* decrease or scaling down of this distance over time in comparison to ever-expanding objects. The variable, ***n***, is the number of seconds that have passed since the original distance was measured between the two objects, and the value shown for X_A is the same universal atomic expansion rate calculated earlier - which never changes. This is the equation for all falling objects and all objects floating in space as they effectively approach each other due to their mutual atomic expansion (an effect currently thought to be due to Newton's attracting gravitational force).

"We can now see that there are sizable differences between the equation of Newton's *Law of Universal Gravitation* and the *Atomic Expansion Equation*. In *Expansion Theory* the 'attraction' between objects actually results from the objects expanding, so the resulting equation is based only on the size of the expanding objects - there is no mass and no attracting force as in Newton's equation. Another obvious difference is that Newton's equation states that the gravitational force diminishes with the square of the distance between objects, yet no such distance-squared term appears in the bottom portion of the *Atomic Expansion Equation*."

Mark McCutcheon, *The Final Theory*, chapter 2

3.7 Second type of growth and shrinkage of matter - The Crossover Effect - Static electricity

In Atomic Expansion Theory, we finally have an answer to this phenomenon of static electricity, and it is not the same as gravity. Everything can now be explained through the simple expansion of electrons and subatomic particles, which are now solely made of electrons; and the expansion of atoms and any object, which are all made of atoms linked together into molecules. There are no more charges in Atomic Expansion Theory. But then, how can particles attract and repel each other? Let's study Benjamin Franklin's electrical attraction experiment in figure 5:

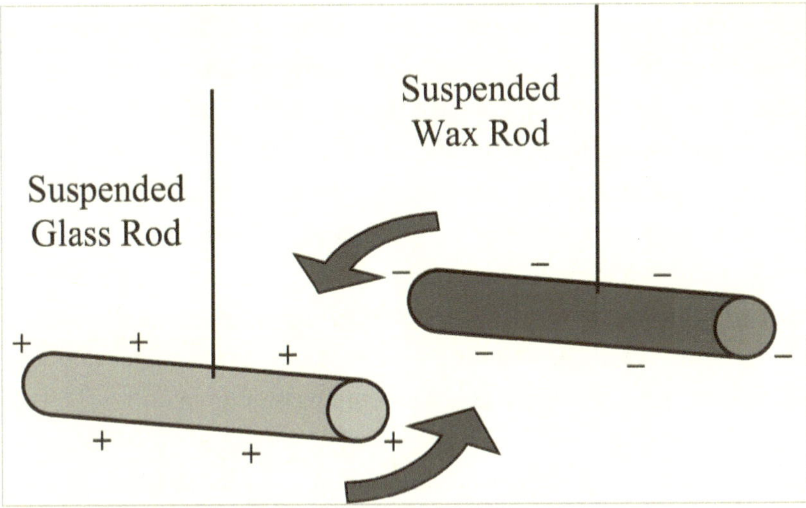

Figure 5 (4-8) Franklin's electrical attraction experiment

In the Benjamin Franklin's "Electric Charge" experiment, he used two rods, one of glass and one of plastic (or wax in the original experiment). He then rubbed them with silk to make the glass rod positively charged, and the plastic rod negatively charged. And when suspended next to each other, the rods attracted each other. And if he used two glass rods and rub them with silk, they were both positively charged, and when suspended next to each other, they repelled each other.

Put very simply, while Mark McCutcheon explains this over several pages in his book, when you rub a rod of glass with silk, you create a surplus of electrons on the surface of the rod. This cloud of electrons, depending on its density, meaning how many electrons there are and how spaced out they are, will start to externalise out of the subatomic realm at a slow growing rate, to the outside where they start to grow freely and greatly. This is called the crossover effect.

Please note that in the crossover effect, we use the terms growing and shrinking instead of expansion, to make the distinction with the inner expansion of the electrons, which is constant behind the scenes (first type of expansion of matter). It is unrelated to electrons growing or shrinking when moving from the subatomic realm to the outside, or from the outside back into the subatomic realm (second type of growth and shrinkage of matter).

I must point out here that for me, the inner expansion rate of the electron is the same as the one of the atom. It would explain why a static field can remain immobile, while everything else around expands at a constant rate. However, for Mark, the expansion rate of the electron and all subatomic particles, X_S, is much greater than the expansion rate of the atom, X_A, and remains unknown.

This growing and shrinking rate of electrons is variable, while still in contact with the subatomic realm, and can even come to a stop, as we will see shortly. But once detached and left to grow freely into space, the expansion rate X_S of subatomic particles and electrons becomes very high, which, while unknown, plays a role in defining the speed of light. This said, I define the speed of light differently, as I will explain later.

The significant difference between our diverging opinion, is that for Mark, the electron is always expanding at that rate X_S, whether it is inside or outside the atom, and whether it is growing out or shrinking back within a static field. And why electrons don't burst out from the atom, or from clouds of electrons in static fields, in their extreme inner expansion, is because the inside of the atom is a different foreign dimension not part of our reality. Our reality would be created by the atoms, and more specifically by the space created between the atoms.

The fact remains that, in a static electric field, electrons are outside the atom, and outside that foreign dimension, and so should immediately start to expand greatly at that inner expansion rate X_S. In which case, the field could not remain static. The electrons would grow extremely fast compared with the atoms within the rods, and the distance could not remain the same. Therefore, I see no need for a foreign dimension within the atom, as soon as we consider the expansion rate of the subatomic particles and of the electrons, as being the same as the one of the atoms and the molecules.

As to how electrons can leave the atoms, if they don't have a much higher inner expansion rate than the atoms, I feel that their growth when excited must be sufficient to kick them out of the atoms, whether they are bouncing, as Mark says, or orbiting, as I believe.

To return to our discussion, when you have two charged rods of glass with clouds of growing electrons of the same density, their expanding clouds of electrons will meet in the middle, growing against each other and repelling the rods. If left untouched, they could repel indefinitely, as seen in figure 6.

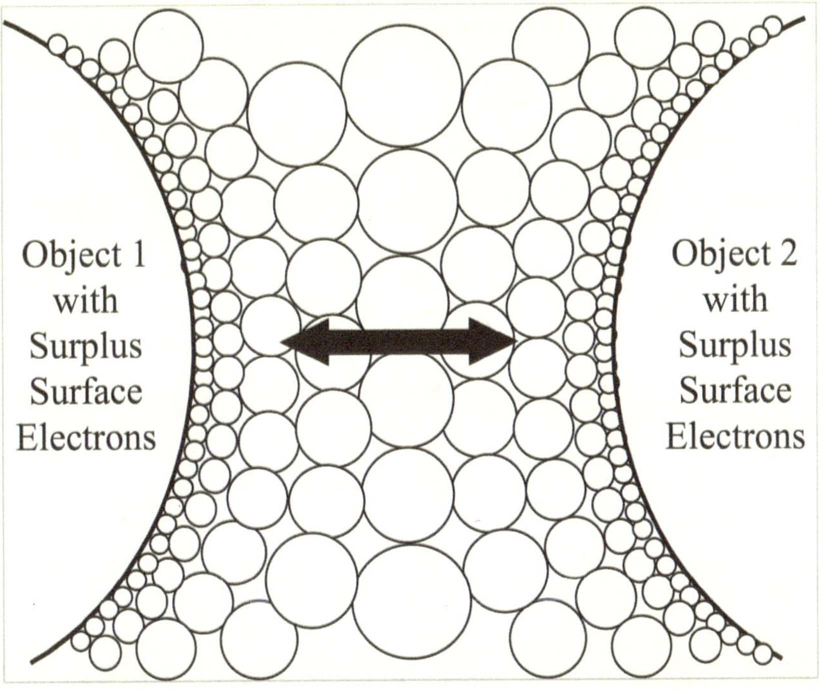

Figure 6 (4-9) Freely expanding surplus electrons cause repulsion

Both rods are pushed apart by a bridge of growing electrons in the form of a cloud, which is the definition of a magnetic field (electron clouds). There is still here an underlying expansion of the electrons, but it is unseen as we also expand. If it were not the case, a static field could not remain static or immobile for long, since everything else around us would overtake the field in size, and it would soon disappear from view.

When you present the glass rod with a denser cloud of expanding electrons on its surface, to a plastic rod with a less dense cloud of electrons on its surface, then the two clouds of electrons will meet closer to the object that has a less dense cloud of electrons on its surface.

When the expanding electrons on the glass rod meet with those of the less dense cloud of expanding electrons from the plastic rod, they start to shrink back into the subatomic realm, so the entire bridge of electrons seeks some sort of balance and equal size, which is the crossover effect. Then the objects are suddenly attracted to each other, as seen in figure 7.

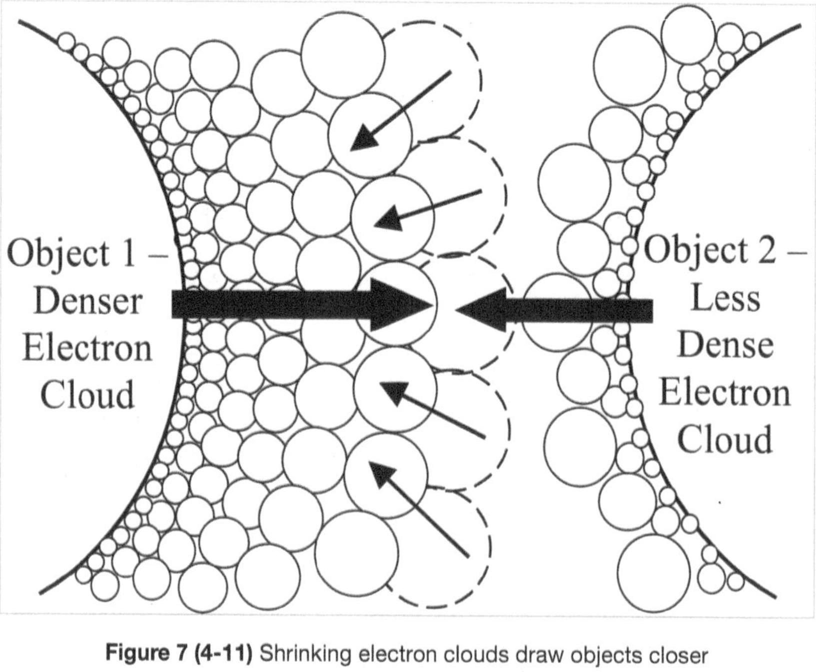

Figure 7 (4-11) Shrinking electron clouds draw objects closer

You can move away the rods and move them back closer together, and the electrons in the bridge will grow and shrink back continually. But once the objects touch, that is it, you will see sparks and hear crackling. The bridge of electrons between the two objects will balance out and return to the subatomic realm, and some of these electrons will escape (sparks and crackling), and both objects will become neutrally charged, no longer attracting each other.

It is different when the two repelling glass rods with clouds of growing electrons of the same density finally touch. Here there is nothing to balance out, so they pretty much keep their expanding clouds of electrons or "repelling force".

It is worth noting that this is why an object is no longer charged positively or negatively. Instead, it is all due to the behaviour of the electrons as growing and shrinking entities, pushing and pulling against each other in their search for balance. Here we have rods with clouds of a surplus of electrons of differing or equal density, and the crossover effect of the electrons between the subatomic world, where they grow slowly and regularly, and once freed outside the subatomic realm, where they grow freely and greatly.

Electrons grow slowly while they are in the subatomic realm, and they can stop growing altogether when two fronts of expanding electrons meet halfway between two rods of glass. They can start growing freely a great deal, once freed completely from the subatomic realm, to become light and heat for example, or magnetic fields. At that point, once completely freed from the subatomic realm, they are no longer growing, instead they are expanding freely into space. And they can also shrink back to the subatomic realm, when attracted back to it, because of a depletion of electrons in a certain area, or the attraction of a smaller or less dense cloud of electrons. All of this explains the attracting and the repelling forces in electromagnetism.

This is the crossover effect explaining the different behaviours of the electrons while in the subatomic realm (inside or just outside the atoms), or when outside the subatomic realm altogether, freely expanding into space, or in between: growing out, halting their growth, or shrinking back in.

3.8 Magnetism and Electricity

Mark McCutcheon has an entire chapter in his book *The Final Theory* to explain magnetism, electromagnetism and electricity, electric circuits, batteries, radio waves, capacitors, and even how old TVs used to work. Static electricity discussed in the previous section is just a small part of that chapter 4 titled *Rethinking the Atom and its Forces.* And it even continues in chapter 5 named *Rethinking Energy.*

The most important part, is that there are no more electric charges, or positive and negative electrons or particles in Atomic Expansion Theory. Mark McCutcheon has shown how the simple expansion, growth and shrinkage of the electrons pushing and pulling against each other in the crossover effect, by moving to and from the subatomic realm, can justify all observed phenomena in physics, including why we thought subatomic particles were charged. I believe the discussion about static electricity in the previous section shows an excellent example of that.

Magnetism, very much like an electric field, is growing electrons outside the atomic realm in a crossover effect, which expand greatly in clouds of electrons. Their movement is explained as moving from locations with a surplus of electrons, to other locations depleted of them, in order for the electric circuit or system to reach a balance of electrons, both in size and numbers.

Using a magnet on an iron bar, which is a conducting material, simply moves the electrons within the bar from one end to the other, creating a surplus of electrons on one side, the positive pole, and a depletion on the other, the negative pole. And the movement of pushing and pulling electrons within clouds, surrounding the iron bar, equalises their sizes and numbers, by moving from where there is a surplus to where there is a depletion. This is what creates the attracting and repelling forces of magnets. And for electromagnets, the difference is that the magnet, or clouds of electrons, are artificially produced through a flow of electricity.

Electricity is simply expanding electrons just outside the atomic realm, pushing each other around wires, which is called current flow. They're not allowed to grow greatly, since they are not freed from the subatomic realm. The movement of the electrons goes from the power source, filled with electrons, to an area depleted of them, like the negative side of a battery, or the ground or Earth for the power in our home.

The flow does not impart any energy to any device connected to it, but affects the atomic material of these appliances. For example, through kinetic energy, exciting the atoms, etc., to produce whatever effect that is required from the device. It can manifest through clouds of electrons as in magnetic fields, or clusters of electrons as heat and light. And when the power oscillates back and forth through an antenna, bands of compressed electrons are launched freely into space to become radio waves.

Here is an interesting excerpt from the beginning of chapter 5, *Rethinking Energy,* of *The Final Theory,* which describes what occurs when you turn on the light. I believe this summarises very well what is electricity according to Atomic Expansion Theory:

> "The True Nature of Light Revealed
>
> "New Idea: Light is not composed of waves or photons of 'energy,' but of clusters of expanding electrons.
>
> "In Expansion Theory, as shown in the previous chapter, electricity is the flow of expanding electrons along a wire as their expansion pressure pushes them along while they are drawn back into the subatomic realm at the other end of the circuit. As the electrons enter the light bulb filament they sink into its volume and cause its atoms to vibrate as they move through. So far, this is similar to the standard explanation of how the filament heats up internally (atomic vibration), but it does not yet address the phenomenon of radiant heat and light.

"The next phase is where the two theories have a dramatic departure from each other. Standard Theory claims that energy is extracted from the electron stream to produce radiant heat and light energy in a conversion process that is not physically understood and cannot even be validated by a corresponding energy loss in the stream leaving the light bulb. Expansion Theory, however, states that it is not 'energy' that is being extracted from the electron stream to create light, but rather, it is the electrons themselves that are being ejected into space.

"According to Expansion Theory, the vibrating molecules of the filament impede the flow of electrons passing through while also agitating them, causing them to dissociate from the atoms and gather into pools where their rapidly building expansion pressure pushes them out into space. This describes a vast sea of expanding electron clusters of all sizes across the entire surface of the filament, continually pushing each other outward – a phenomenon we know as radiating heat and light (Fig. 5-1 [8]).

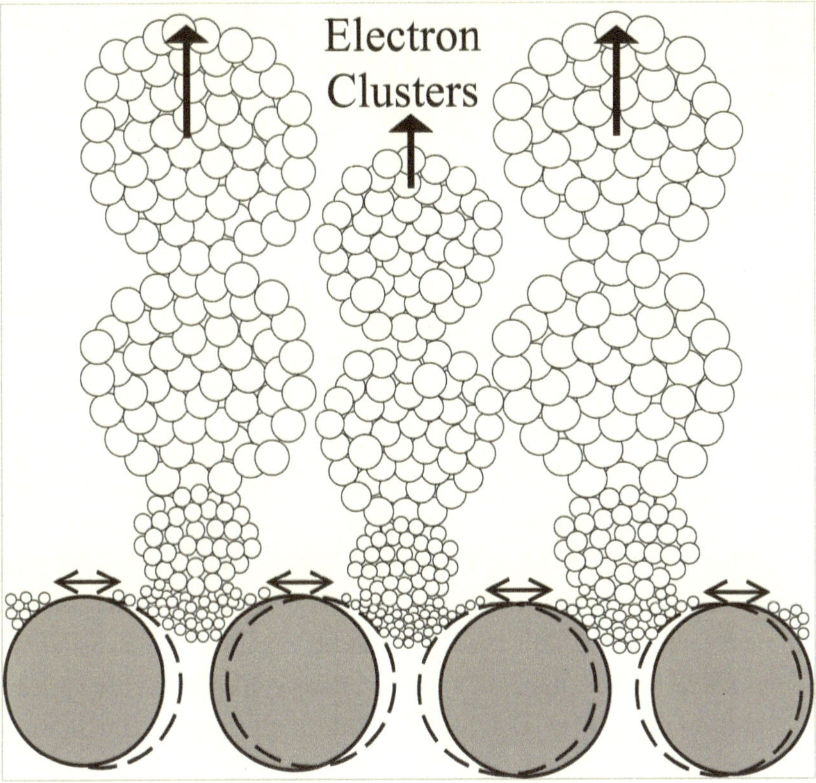

Figure 8 (5-1) Vibrating atoms eject electron clusters (heat and light)

"Expansion Theory claims that heat and light are not photons or waves of 'pure energy' mysteriously produced by electrons flowing through a resistor (the light bulb filament), but rather, they are physical clusters of expanding electrons pushing one another through space.

"The initial heat that radiates from the filament before it begins to glow white-hot is the larger electron clusters, caused by the larger pools of less agitated electrons that would be able to gather while the atoms vibrate relatively slowly and expansion pressure builds more slowly before ejecting clusters. As the atoms vibrate faster and faster the increased agitation creates ever-smaller pools of expanding electrons that fill rapidly and eject smaller clusters more frequently – a phenomenon known as light. This is the reason radiant heat is considered to have a longer wavelength than light; heat is composed of larger clusters of electrons."

Mark McCutcheon, *The Final Theory,* chapter 5

3.9 Energy

Light and heat are clusters of electrons expanding greatly and freely outside the subatomic realm. Their speed, or the speed of light, depends on their constant expansion rate. Everything is essentially made of electrons, including all radiant forms of energy. Energy including light is made of matter instead of photons, meaning light is made of different sizes of clusters of expanding electrons.

Radiant energy is still physical in nature, consequently, anything made of light or energy, including spirits or ghosts, is still part of our physical reality, it is still made of matter. Although I believe such beings are likely also made of atoms and molecules, and only appear as light to us due to the higher vibrational rate of the matter composing them.

The difference between a positively and a negatively charged atom, or an ion, depends respectively on if the atom loses or gains electrons. Then of course, they either repel or attract each other depending on if they have missing electrons, or a surplus of them, compared with the number of protons, in order to reach a balance.

Ions are atoms or molecules in which the total number of electrons is not equal to the total number of protons, giving the atoms or molecules a net positive or negative electrical charge. Two ions with the same number of electrons would repel each other, as their electrons are bouncing off their nuclei. If one atom has less electrons while the other has more, then they will attract each other. An electron will more readily then bounce from the nucleus of one atom, to the nucleus of another atom, in search for balance. Again, this is the origin of chemical bonds in Atomic Expansion Theory.

Therefore, in Atomic Expansion Theory there is no positively or negatively charged particles, but there are equivalent concepts to explain chemical bonds and other charged particles phenomena, such as how two clouds of electrons will repel or attract each other, as mentioned before in the sections about static electricity and magnetism.

See below the physical descriptions of all forms of energy, taken from *The Final Theory,* Table 5-1 *Today's energy terms according to Expansion Theory*. Only the last one, kinetic energy, is considered to be a real form of energy. This is when particles and objects gather speed through the geometry of expansion, resulting from the natural orbit effect and relative motion, and have a real and greater force of impact when they hit something. Orbits and motion will be explored in the next section, and it explains why the electrons move from where there is a surplus to where there is a depletion, as it is all part of the geometry of expansion.

Gravitational Energy

An *effective* attraction between all objects due to their ongoing expansion as objects composed of continually expanding atoms.

Strong Nuclear Force Energy

The natural cohesion of protons and neutrons in the nucleus of an atom due to their tremendous ongoing subatomic expansion against each other.

Electric Charge Energy

Attracting or repelling forces caused by a *crossover effect* of externalised expanding subatomic particles acting between subatomic and atomic realms.

Chemical Bond Energy

A manifestation of the *crossover effect* between subatomic and atomic realms, occurring between individual atoms rather than overall objects.

Magnetic Energy

Clouds of expanding electrons surrounding conductive objects, attracting or repelling via the *crossover effect* between subatomic / atomic realms.

Electromagnetic Energy

Bands or clusters of freely expanding electrons that continually push one another through space due to their ongoing inner subatomic expansion.

Kinetic Energy

The *apparent* absolute energy of motion "possessed" by objects, but which is actually only a purely *relative* motion effect between objects.

3.10 Motion and orbits

Motion of everything in the universe is relative and entirely geometry based. There is no longer any kind of gravitational force acting at a distance emanating from matter, which attracts objects together, as stated by Isaac Newton. Gravity is the changing distance between expanding objects, and this is what has been perceived as an attractive force acting at a distance. Like it is in Einstein's General Theory of Relativity, which is also geometry based, there is never any kind of resistance on any object. They simply move or enter orbits because of the geometry of expansion, as will be seen in the next figures.

Here are two figures showing what is happening when two objects pass each other in straight lines in space, and how it leads to a natural orbit effect. This is while considering that they and we are expanding proportionally with everything else, and therefore we cannot see this expansion:

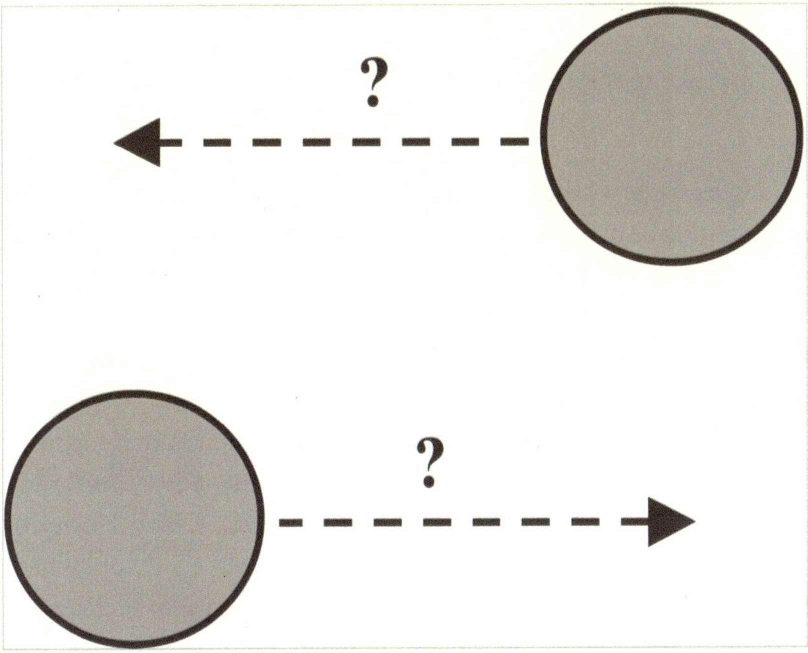

Figure 9 (3-5) One object speeding past another

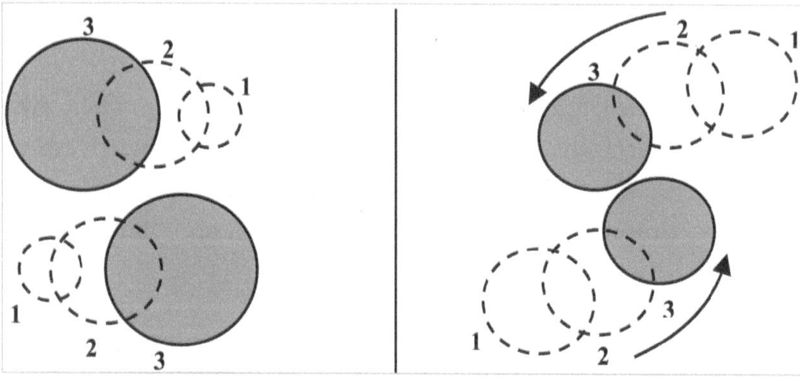

Figure 10 (3-6) Concept (left) and result (right) of the Natural Orbit Effect

Figure 9 shows the relativity of motion, with the question marks representing our inability to assess if both objects are moving, or just one, and if so which one, and what their respective speed actually is from the point of view of each other. If there are no other viewpoints from which it could be assessed, then there is no way to know, and all options are actually possible.

For Mark McCutcheon, the relativity of motion is more akin to Galileo's Relativity, instead of Einstein's Special Relativity. As for my point of view, I will discuss in the next chapters a new Theory of Universal Relativity concerning relative motion. It is a theory where objects don't move, in a universe without real space or distance. Objects simply expand or shrink from our point of view, as they come towards us or move away from us.

Figure 9 also shows the perspective if there was no expansion or gravity in the universe, the two objects would then simply continue on their straight lines unaffected. This perspective does not exist in nature, neither behind the scenes nor in our resulting reality, because objects constantly expand behind the scenes.

In figure 10 on the left, the two objects are expanding in their straight line trajectories, but this expansion is unseen by us because everything expands proportionally at the same rate, including our instruments of observation, our rulers and ourselves. What we see and experience instead, is what is on the right: the distance reducing between non expanding objects, creating the Natural Orbit Effect of all moons and planets.

Effectively, in the resulting reality following from what is happening behind the scenes, nothing could ever move in straight lines in the universe, or ever stay at rest, the opposite of Newton's First Law of Motion.

Still in figure 10 on the left, we find what Mark McCutcheon came to call the God's viewpoint, which is simply an expression, which is a standpoint showing what someone outside the universe would see, if they were not part of our universe and not subject to expansion. From that vantage point, they could then see the expansion in our universe.

In the same image, the expanding and larger objects have now passed each other without entering into a collision. As the objects continue on their respective straight line trajectories, if they are not going fast enough, they would be unable to escape each other in their expansion, and would eventually join together. If they are going fast enough, they will be able to counteract their mutual expansion, and keep at a certain distance from each other, and this would then result in a stable orbit as seen on the right side, where the expansion is unseen.

If the objects were going at the right speed, the gap between them might never change, since expansion each second would reduce the growing distance between them, in just the right amount to match their respective ongoing coasting speed. It could then appear as if they were not moving away from each other, while expanding. But if they were not moving, they would soon join together.

From the God's viewpoint, it is relatively easy for two objects, expanding at a constant rate, to keep at a certain distance forever, while simply continuing on their original constant speed. Remember that behind the scenes there is no gravity, there is no orbit, and there is no distance reduction in an acceleration. There are just expanding objects immobile or moving in all directions, sometimes hitting each other.

For example, in figure 11 we see what happens when an object is passing by the Earth at the same rotational speed of the Earth, and the geometry of expansion locks it up into a geostationary orbit. Once again, you have on the left what happens behind the scenes, where you can see the expansion, and the result in our reality once you can no longer see the expansion:

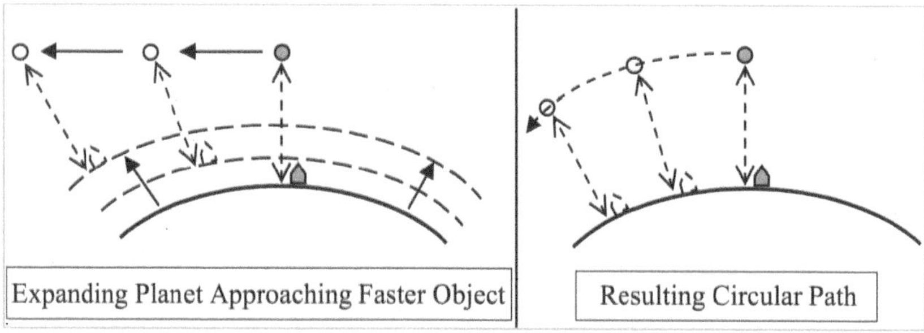

Figure 11 (3-9) A faster-moving object produces a circular orbit

Each new second, the geometry is identical to the previous second, because the object is moving at the exact speed required to compensate for the expansion of the planet and its rotation. As a consequence, the object in orbit remains in the same position over the house, as if fixed there in the sky.

To calculate orbits in Atomic Expansion Theory, we use Kepler's three purely geometric empirical equations of planetary motion, which involves no gravitational force nor mass. The Newtonian Orbit Equation has been replaced by the Geometric Orbit Equation, which Mark McCutcheon says is a previously unrecognised, purely geometric equation, embedded in Newton's equation, and extracted from patterns in the standard astronomical data available at the time of Kepler and Newton. There is more detail about the Geometric Orbit Equation in chapter 1 of *The Final Theory*.

Mark McCutcheon has an entire chapter to explain orbits in his book *The Final Theory,* chapter 3 titled *Rethinking Our Celestial*

Observations, but hopefully this summary gave you a good idea of how expansion can lead to a natural orbit effect, and what is meant by orbits following the geometry of expansion.

3.11 Behind the scenes - Four different perspectives required to explain orbits and gravity

3.11a The first perspective - The God's viewpoint with expansion - And can we change the expansion rate of matter?

To understand orbits within Atomic Expansion Theory, it is essential to visualise three different perspectives, showing all the mechanisms behind the scenes, that lead to what we see in our reality. There is a fourth perspective I will discuss at the end, but for now let's concentrate on the first three. As mentioned earlier, the first one is called the God's viewpoint, and it is the perspective we would see if we were outside the universe, and were not expanding ourselves, while actually being able to see the expansion.

The God's viewpoint is entirely made of expanding objects going in all directions, where there is no possibility of any kind of orbit forming naturally, since orbits are solely the result of the relative motion of the geometry of expansion from our reality. An effect that results from the fact that there is no expansion anymore, and yet all the consequences of this hidden expansion are there. Only in our reality would objects enter into orbits, or accelerate, or slow down due to the geometry of expansion. Unlike in our reality, from the God's viewpoint, objects don't change speed or direction of motion, unless they hit other objects.

Mark McCutcheon's says that this viewpoint does not exist for us and never will, because we will always expand proportionally and relatively along with everything else. Therefore, no one but a God, not part of our expanding universe, could ever witness that expansion. He even goes so far as stating that there is no expansion in our resulting reality, and we certainly cannot see it when we look around us.

There is only a reality where we can see the after effects of this hidden expansion, such as the distance between objects reducing in an acceleration, or a satellite gaining an unusual acceleration from being sent into a partial orbit around a planet. But for all intents and purposes, that other reality behind the scenes, where objects double in size every 19 minutes, does not really exist. Consequently, we don't need to worry about all that expansion behind the scenes, because in our resulting reality, nothing is really expanding. I agree with this point of view, as one way of looking at it.

However, the fact that we cannot see the expansion, does not mean that it's not happening. Although no one will ever see that reality behind the scenes, it is what is underlying our reality. And I prefer to look at it as a reality that truly exists, as part of our reality explaining what we see, which is just seen from a different angle. This said, it does not matter whether we see it as existing or not, it is what is underlying our reality at any rate. And who knows, if someone in time manages to reduce the constant rate of the inner expansion of all their particles, although Mark McCutcheon believes this to be impossible, they might actually find themselves seeing our reality from the God's viewpoint.

It is important to note that all the orbits, and the coherence of the electrons, atoms and molecular structures, can only be maintained if everything is constantly expanding. Therefore, you would not want to stop the inner expansion of matter. But who knows about increasing or decreasing the expansion rate of matter?

It would be difficult, because that first type of expansion of matter, in my opinion, must be due to the inner expansion of the matter composing our particles, at yet another very small scale universe, smaller than our very small where we find our familiar particles. And the inner expansion of that scale of even smaller particles, must be due to the inner expansion of the smaller particles composing them, at yet another smaller scale. I believe it is very much like the structure of fractals.

Perhaps we might still be able to find a way to decrease or increase the inner expansion rate of matter, although I have no idea how this could be achieved. It would immediately lead to an object suddenly expanding enormously, compared with everything else, possibly larger than the planet or the solar system. Or quickly reducing in size to a scale smaller than our very small, as things around us behind the scenes would continue to double in size every 19 minutes. Even if somehow we could manipulate the expansion rate of matter, it could be extremely dangerous.

This a relative universe, and we don't know everything yet about what is possible or not, and what the effects might be. Increasing or decreasing the constant expansion rate of matter for a very short time, might lead to manageable and interesting results. I have written the beginning of a science fiction novel in French called *Made in Québec, La Renaissance of the Nouvelle-France,* where I use small but intricate models of houses and castles, and instantly expand them to larger than their usual size, solving the housing crisis. Or instead of multiplying fish and bread, I simply expand them to a much greater size, in order to feed the masses. The way I made the technology to expand and shrink objects in the novel, is by some cloud of gas causing a chemical reaction, that somehow affects the Universal Atomic Expansion Rate of matter. Such a thing is very unlikely to work, but it is science fiction.

If I had to think something up once again, for another sci-fi project, I would invent a device like a large capacitor or crystal, capable of amplifying and directing thoughts into objects I want to expand or shrink. Very much like Ralph Ring and Otis T. Carr did with the flying saucers they built, which I discussed in chapter 2. It makes sense that will and thoughts should be able to affect anything in our reality, if somehow we live in a computer simulation, or a virtual reality of some sort, or a holographic universe. Because then, we would be living in a psychological or psychophysical reality of some kind, where thoughts can easily affect anything.

I must point out that what Ring and Carr were manipulating, was the third type of growth and shrinkage of matter, and not the expansion rate of matter, which is the first type of expansion. Objects can grow and shrink using that third type, but then they move to other dimensions, outside of our reality. You could always expand the wine that way, in order to turn the masses into drunk raving lunatics, but it wouldn't help much if only Jesus and the aliens living in the fourth dimensional world could drink the wine.

To get back to our discussion, there is also not much that can be gathered from the God's viewpoint to explain what we see. For example, there is no way to know at what speed these expanding objects are going at behind the scenes, or their direction of motion. What is important, is to understand the general idea that once the expansion is hidden, all sorts of geometrical relative effects occur, and it explains what we see.

Someone clever one day might find a way to calculate exactly what is happening in the God's viewpoint. And maybe it would be useful to know the real speed and direction of motion of these expanding objects behind the scenes, such as the Earth, the Moon and the Sun. It might help to explain all the trajectories of the orbits.

3.11b The second perspective - Our resulting reality without expansion

The second viewpoint is our resulting reality of gravity and orbits, but of relatively stable orbits, where we cannot see any enlargements of the orbits, or the expansion of the solar system. For example, the Moon looks like it is always keeping at relatively the same distance from Earth. And all the planets' orbital rings also appear to keep their relative distance from the Sun, and from each other, despite their inner expansion. This is a very neat and orderly reality, but it is not the whole story.

3.11c The third perspective - Expansion re-established after the relative effects - And how orbits enlarge in spirals and naturally accelerate objects

There is a third viewpoint, which comes into view only after our resulting reality has come to exist. It is a viewpoint that is a consequence of the second perspective, our resulting reality, which explains the relative effects that we witness. These effects are only explainable if we re-establish the expansion of matter after the facts. Although I also call this third viewpoint as being behind the scenes, for convenience, it actually results from our reality, and it is quite distinct from the God's viewpoint.

We could debate if this third reality exists for real, or if it is only there to help us visualise what is really happening. However, we must remember that these are solely three different points of observation, or perspectives, of a same unique universe and reality. In the first one we see it with the expansion, in the second one we see it without the expansion, in the third one we see it with the expansion after all the relative effects.

For example, the solar system is expanding, and for that to happen, the planets' orbital rings around the Sun must be enlarging in spiralling orbits. But for us, these orbits appear relatively stable, without any kind of enlargement that we can see. So how does the solar system expand? Now, this is a very key question, as even that expansion we cannot see.

This third viewpoint is clearly required, because the solar system and all the planets' orbital rings are expanding. And this third perspective is far from the God's viewpoint, where objects are going in all directions, perhaps not even moving at all but simply expanding, and where no orbits are ever possible, all the while creating motion in our resulting reality.

All three perspectives are required to understand orbits and any hidden effect, such as a geometrical acceleration as witnessed in the slingshot effect. Behind the scenes, from the God's viewpoint, as mentioned, objects don't have to accelerate away from each other to counteract their mutual expansion, they can simply move away from each other at their constant speed. And yet, in our resulting reality, these objects undergo a real acceleration, due to the geometry of expansion, resulting from the fact that suddenly there is no more expansion, and yet distances still reduce between everything.

For example, the Moon is probably moving away from Earth in a straight line from the God's viewpoint. Since no orbits are possible there, there is no relativity of motion. Objects expand and may or may not be moving at all. Objects hit each other or move away from each other, in all directions, possibly curving, but nothing to do with any natural orbit effect. Then, once we no longer see the expansion in our resulting reality, as can be seen in figure 10 above, the Moon is actually in a relatively stable orbit around the Earth, more or less always keeping the same distance, following its slight elliptical orbit.

And this is when the third viewpoint comes into play. Once our resulting reality is established, the only way we can explain what we see, is to re-introduce the expansion after the facts. Suddenly the Moon and the Earth are actually expanding, their orbits around each other are enlarging in spirals, and this is how they can counteract their mutual expansion. Once again, this is not to say that this reality exists, but it is certainly a perspective that is helpful to visualise what is happening.

Is it not strange that when we wish to put an object into orbit, we accelerate it to a certain speed and height, then simply let go? And that satellite will continue on its course around the Earth forever, or certainly for many years? What is happening here? The object does not require a mean of propulsion, does not need to accelerate or gain more speed, simply the geometry of expansion keeps it where it is in the sky. So what is happening behind the scenes to explain this?

If we re-establish expansion at this point, the third perspective shows us that any orbit accelerates the objects involved. And not only that, but the speeds of the objects gradually accelerate in time, leading to enlarging spiralling orbits, instead of the circular and elliptical ones we are familiar with.

Gravity reduces the distance between all objects in an acceleration, and yet, a satellite in orbit not only counteracts that gravity, it even accelerates, otherwise there would not be an orbit. The acceleration comes from a slingshot effect from the relative motion of the geometry of expansion, which we find within all orbits. This is what counteracts the acceleration of the distance reduction due to gravity. Gravity has to be counteracted within all orbits, and basically this means the constant expansion has to be counteracted, otherwise all objects would eventually hit each other.

It is easier to see how orbits create a natural acceleration of the objects or particles, when we consider elliptical orbits, because the same must apply to the circular orbits as well. As seen in figure 12, most orbits are elliptical or oval shaped in nature, instead of perfect circles:

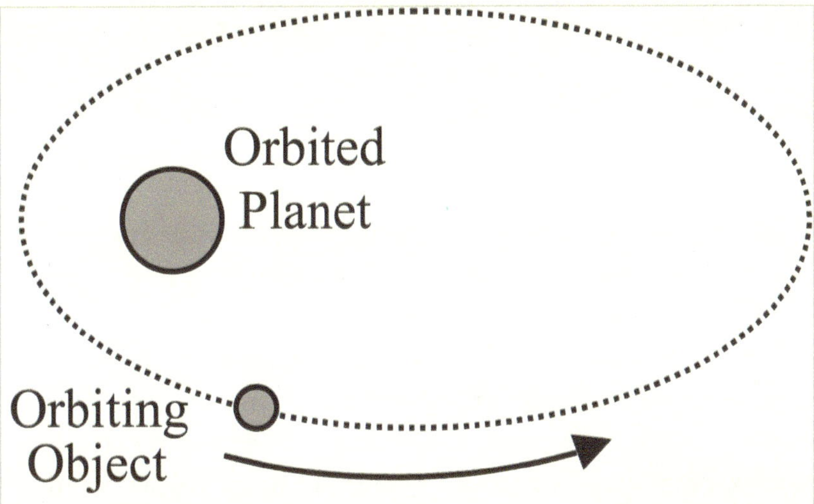

Figure 12 (3-12) Object swings around planet and away in elliptical orbit

From the third perspective, when Earth reaches its apogee on the right, at the other end of its elliptical orbit around the Sun, in its expansion the Sun is catching up with the Earth. At this point, it is like an object falling back to Earth. The Earth starts its return towards the Sun, and the distance reduction between them accelerates, as per the Atomic Expansion Equation to calculate gravity. This acceleration gives the Earth the swing it needs to go around the Sun, and escape once again the expansion of the Sun. And then the Earth goes away from the Sun on its elliptical orbit, slowing down again as it reaches its apogee at the other end.

Now, it may appear as if gravity accelerates the Earth towards the Sun, but then this is cancelled by the Earth going away once again on its orbit, decelerating while the Sun is catching up with it in its expansion. And since it all balances out nicely, overall, there is no extra acceleration that could justify an enlargement of the orbit behind the scenes, or a natural acceleration of the Earth and the Sun.

However, this acceleration is like a slingshot effect of a partial orbit around a planet. Sometimes NASA will have a probe orbiting a planet several times to accelerate it further, which is a clear indication that a simple orbit accelerates objects. When we study the slingshot effect, we can see that a probe sent into a partial orbit around a planet, will experience a net gain in acceleration from its initial speed. This is the whole point of a slingshot effect, and it is entirely happening as a geometrical relative effect due to the expansion behind the scenes. The object leaves with much added speed compared with the speed it was initially going at, and this is a real effect measurable in our resulting reality.

From the third perspective, this acceleration causes the enlargement of the orbit of the Earth around the Sun. This explains why we need to accelerate a satellite into its orbit, but once it is there at the right altitude, the geometry of expansion alone will continue to accelerate that satellite, keeping it into its orbit, despite all the expansion and gravity.

It also demonstrates why two objects from the God's viewpoint, don't need to be accelerating away from each other, in order for the natural orbit effect to accelerate them away from each other in our resulting reality. And this is also how the solar system is expanding, and any other object that has orbiting objects or particles, such as galaxies, atoms and electrons. But of course, we will never witness that expansion, we can only witness the resulting after effects.

To summarise, the third viewpoint of an orbit, would be showing the Sun and all the planets expanding, and their true enlarging spiral trajectories that would lead to the overall expansion of the solar system. It is showing us what is really happening, in a way that the God's viewpoint cannot.

In our reality, the speed of the Earth and the Moon does not change overall, beyond the changing speed during their elliptical orbit. And their orbits are stable, identical at every cycle. It is true that the Moon is moving away from Earth a little bit more every year, but this just means that the orbit is not perfect. The Moon is moving away from Earth in the God's viewpoint, a little bit faster than is required for a stable orbit.

From the third perspective however, not only the orbits are enlarging in spirals, but all the objects are accelerating. And so, in time, they are constantly going faster, which means vibrating faster. This becomes important when we discuss other dimensions in the next chapters, because it means we are constantly moving into higher dimensions, where everything is moving faster.

3.11d Are orbits in the third perspective enlarging gradually or exponentially?

At the end of this chapter, there are links to orbit simulations and discussions about the concepts of Atomic Expansion Theory on YouTube. I've been told that such simulations of enlarging orbits, only work if the orbits enlarge exponentially, following the Golden Ratio Phi and the Fibonacci sequence of numbers.

Although this is a seductive idea, where the shape of expansion and all orbits from the third perspective, could follow the Golden Ratio, and explain why it is found everywhere in nature, I somehow feel the spiralling orbits are not enlarging exponentially. The reason why, is that it would lead to an exponential expansion of solar systems, galaxies, atoms and electrons. And we know they expand at a constant rate.

I think there is confusion here when dealing with accelerations, because by definition they have an exponential compounding element to them. However, an accelerating car does not suddenly multiply its speed exponentially, it is a gradual acceleration, so an orbit could equally simply accelerate progressively.

There are many factors to take into consideration here. An orbit results from the geometry of expansion, and, as just demonstrated with the elliptical orbit above, there is an acceleration, like in gravity, while the Earth is coming back towards the Sun. Then, there is a slingshot effect, just like in partial orbits, that boosts that acceleration of the Earth around the Sun. But then, the Earth decelerates as the Sun's expansion catches up with it. There is an additional net gain in speed following each orbit, but is this gain in speed truly exponential, leading to exponentially enlarging orbits?

When balancing all these facts, I believe that in the end, the enlargement of orbits everywhere is gradually increasing without doing so exponentially, resulting in the overall expansion of all objects at a constant rate, instead of an accelerating exponential one. If orbit simulations truly require an exponential enlargement of the orbits, then some more thinking will be required to explain what is going on, because objects cannot expand exponentially, nothing would work in the universe if they were. So a balance between relative effects must be reached somewhere to prevent the enlargement of the orbits to expand exponentially.

If further thinking is required, then the person behind the YouTube channel *Life, Everything And The Universe,* is working on this from the point of view of Atomic Expansion Theory (the links are at the end of this chapter). He is investigating the Coriolis acceleration and deceleration, tangential and radial velocities or accelerations, fictitious or pseudo forces, and acceleration due to curved surfaces. It is quite possible that these concepts could justify further the geometry of expansion, and show exactly how orbits and the slingshot effect work. It should be interesting to see where it leads.

There is also another point that could warrant investigation. It is a fourth perspective if you wish, one that does not exist but might underlay the God's viewpoint. It is the perspective where we would imagine a universe without any expansion or gravity. The geometry alone would accelerate and decelerate objects due to their curved surfaces, and this, without any expansion. These relative effects would also affect our resulting reality.

In the end, we are not short of ways to explain accelerations and decelerations of objects within orbits and gravity, from the point of view of all the different perspectives. And overall, it must all balance out, so the enlargement of the orbits is not exponential. Even if certain effects, such as the Coriolis effect, would cause the enlargement of the orbits to be exponential. Something else must counteract it, or explain why the overall expansion of the solar system, the galaxy, the electron and the atom is constant.

Another important point here, is that I am only talking about the expansion resulting from the inner expansion of matter, the first type of expansion. The discussion above is not taking into consideration any subsequent second or third type of growth or shrinkage of matter, such as electrons and solar systems, which is on top of that basic constant inner expansion.

For example, as well as constantly expanding due to the inner expansion of matter, our solar system can also variably grow and shrink, depending on how much matter enters or leaves the system, exciting or calming the matter already there. But if Planet X or Nibiru suddenly enters the solar system, don't expect all the orbits to change, grow further in size and move further away from the Sun. If that planet is already orbiting our Sun, then all orbits have already compensated for it. Any change in all the orbits is instant everywhere, because of the geometry of expansion, as explained near the end of section *2.13 A different concept of positively and negatively charged particles in Atomic Expansion Theory.*

Add several new planets not already orbiting our Sun, and the Earth might move away from the Sun sufficiently for us to witness a real change in the weather, while the entire solar system will grow a bit more. It is likely the asteroid belt used to be an extra planet which has been destroyed. The effects on all the orbits, the entire solar system and even beyond, must have been significant.

There is also the growth and shrinkage of atoms and galaxies, and molecules and cosmic molecules to be considered, which leads to matter moving into other dimensions, the third type of growth and shrinkage of matter. I will discuss all this in the next chapters.

3.11e The fourth perspective - Objects passing each other in space if there were no expansion or gravity

Before we even talk about the first perspective, what would happen if objects were to meet in space while there is no expansion or gravity, or any other relative effect? How does the geometry alone affect the distances or anything else?

Underlying all perspectives, there are relative effects that result from the geometry of expansion, but there are also relative effects that may result from the geometry alone, such as the geometry of curved objects, if there were no expansion or gravity in the universe. These effects, such as accelerations and decelerations of objects, would be found in our resulting reality, and so they need to be assessed and taken into consideration.

3.12 Slingshot effect and other gravitational anomalies

Using the Atomic Expansion Equation to measure gravity, we obtain more accurate results than using Newton or Einstein. As mentioned in the previous sections, gravity is a changing distance between expanding objects, measured from surface to surface, nothing to do with their respective mass measured from centre to centre of the objects.

Although there is a correlation between mass and size, it is not always true, as in the case of a large planet that would be made of Styrofoam, or of a lighter material of a lesser density, like gases. These planets could be extremely large, but certainly their mass would be less than a planet made of rocks.

From a distance in space, a planet made of Styrofoam would exhibit the same gravity as one made of rocks. The distance between each of them and an asteroid, would reduce at the same accelerating rate. Once standing on the planet however, the density of that Styrofoam planet would make gravity much less than one made of rocks, because in its expansion, a planet of a lesser density cannot push back as hard as one of a higher density.

There you see where the problem lies when relying on mass to measure gravity and orbits, especially when our calculations of the mass of planets is more like a guessing game. This is why in our space programmes, while using Newton, we must constantly readjust our calculated trajectories, which always run off course. It could also very well explain the *Flyby anomaly,* which you can find out about on Wikipedia.

Although Mark McCutcheon does not believe that galaxies are expanding, as mentioned earlier, he does believe that solar systems are expanding. As just seen, behind the scenes, the true orbits of the planets around the Sun must be enlarging spiral orbits, counteracting their mutual expansion.

There is proof of this, one main proof is the Pioneer Anomaly. The spacecraft Pioneer 10 and 11 have sort of exited the solar system, or are in the process of doing so, but strangely enough they appear to be decelerating. Scientists came up with several feeble explanations for this, which do not stand to scrutiny. Pioneer 10 and 11 are not actually decelerating, instead the solar system and all its planets' orbital rings are expanding towards them. It was not very clear before, as they travelled across the solar system, but as they are coming to the edge of the solar system, it is becoming obvious.

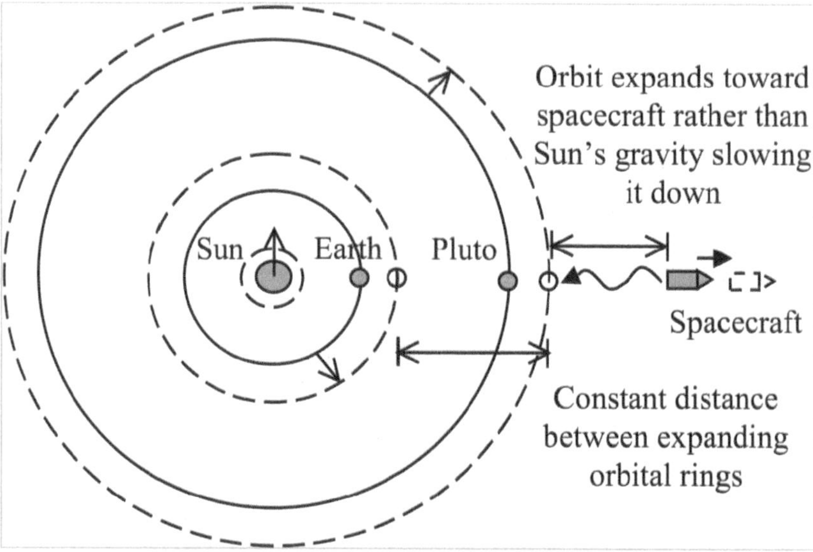

Figure 13 (3-26) Expanding solar system differs from Newtonian gravity

As seen in figure 13, first, the distance decreases between Earth and the probes, as its orbital ring enlarges and the solar system expands. Second, a signal sent from the probes takes less time to reach us, from the time it is sent until it arrives, since in the meantime the Earth is moving towards the signal, as its own orbit around the sun expands. You can read here an excerpt of *The Final Theory* on this subject, available on my website *www.themarginal.com:*

Pioneer Anomaly, Slingshot Effect and Gravitational Inconsistencies Explained

www.themarginal.com/pioneer_anomaly.pdf

There are several other proofs for Atomic Expansion Theory. An article I have included at the end of this present book, called *10.4*

Gravity Breakthrough: Springing into a Gravitational Revolution, is the greatest proof according to Mark McCutcheon. But I wonder about that proof, and also its equivalent, when we drop two balls attached by an elastic band, as described in the first article *10.1 Expansion Theory - Our Best Candidate for a Final Theory of Everything?*

I fear somehow scientists are beginning to deny Einstein's Principle of Equivalence with Newton's theories. At the very least, the *Gravity Breakthrough* article proves that much. People could somehow say, that Einstein can explain that spring and elastic band experiments as well, because his theory of gravity is not equivalent to Newton's after all. Simply because, Einstein's Theory of General Relativity is also geometry based, and no actual force is applied anywhere on any object.

Essentially, there would be no Principle of Equivalence anymore between Einstein and Newton, if they were to say that Einstein can also explain these experiments, and yet, it is unlikely to stop them from saying it. To be honest, it is true that Einstein's Theory of General Relativity is geometry based, and that no force is acting at a distance on these objects, and that probably, Einstein's Principle of Equivalence is incorrect after all. So we need better proofs.

For example, water falling from the sky, would turn into a sphere in both Einstein and McCutcheon, because unlike in Newton, no force is acting on the ball of water at any time. But is the ball free floating in the air while the Earth is expanding towards it, or is the ball free floating, as well as following the curving of spacetime, that leads straight to the ground? Keep this in mind, while watching the links to videos on YouTube at the end of this chapter. Some proofs could easily be questioned, and Einstein's theories easily tweaked, to justify what we see.

This said, there are other proofs. For example, McCutcheon has for the first time calculated the Gravitational Constant G from first principles, derived from his Atomic Expansion Equation. He shows the derivation in his book. So far it has only been calculated through empirical evidence, meaning through observation instead of pure reason alone. Although, I must say, that Mark does not feel this is a great proof. So I guess this too could be twisted around somehow by people determined to prove Atomic Expansion Theory wrong.

However, his entire book is simply proof after proof that all of physics can be explained through the expansion of electrons and atoms, so what more proof is required? Nevertheless, in the next section I present the two best proofs I could think of so far.

This is not your normal run-of-the-mill crackpot theory most theoretical physicists receive on a daily basis, which made them so far refuse to consider Atomic Expansion Theory, as they can't possibly consider every new theory there is on the block. But consider this, how could anyone explain the entire physics in a way that makes complete sense to everyone, without requiring a PhD in Physics, solving all the anomalies, questions and bizarre spooky effects of our actual science, while uniting and explaining the four fundamental forces of nature, if he was not correct?

The Moon does not have any gravitational effect on the Earth. The tides of the oceans are due to an internal wobble of the Earth which coincides with the passing of the Moon, due to the original formation of the Earth-Moon system. Orbits are entirely a geometric phenomenon, due to the geometry of expansion of the relative motion of objects in space.

On that subject, even with Einstein the Moon cannot possibly have any gravitational effect on the tides, since General Relativity is also geometry based, and there is no force acting at a distance there either. Objects are meant to be free floating in space, with no force acting upon them.

This was an important obstacle for Einstein, who desperately tried to explain how the Moon could still affect the ocean tides on Earth, through the curving of spacetime and geodesic deviation. It does not sound very convincing. It might have been better for Einstein to simply state that the tides were due to something else closer to home, there are several other possible explanations that can be found online. If he had done that however, there would not have been an equivalence with Newton's theory, instead, there would have been significant differences.

As it stands now, even with Einstein the Moon can affect the tides, precisely as if there was a force acting at a distance emanating from matter. In which case, Einstein's Principle of Equivalence with Newton still stands, and therefore Einstein cannot be used to justify the spring and elastic band experiments. Because if Newton can't explain it, then Einstein can't either, since their theories are equivalent. Whether it is a force acting at a distance, or geodesic deviations on a curved spacetime, the result is exactly the same for any experiment.

In Atomic Expansion Theory, gravity depends on the size of the objects, not on their mass. Although density still plays a role in gravity, but only while on the ground, and it could not be assessed from orbit. This is why where NASA landed their probes on the Moon, gravity was only one sixth the one of the Earth, instead of the quarter that was expected, since the Moon is about a quarter the size of the Earth.

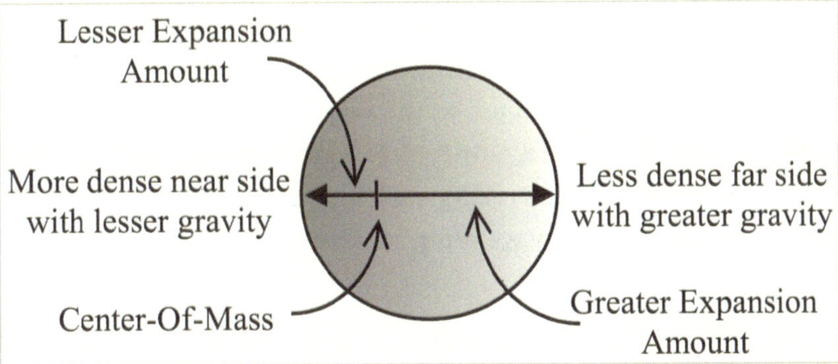

Figure 14 (3-14) The moon's non-uniform density causes differing gravity

As seen in figure 14, the density of the Moon must be unequal, and be denser nearer the Earth, due to the original Earth-Moon formation. It also means that on the dark side of the Moon, gravity should be one third the one of the Earth, and about a quarter near the edge of the visible Moon. Taken overall, gravity around the Moon in average is about a quarter of the one of the Earth. It is a wonder some countries were able to land probes on the Moon while being unaware of this. Other countries have not been so successful, and their probes crashed on the Moon.

Gravity is calculated from surface to surface of objects, not from centre to centre. This clarifies several observed trajectory and gravitational anomalies in space and on the Moon. There are no gravitational forces or gravitational waves acting at a distance, or distortion/warping of a so-called spacetime, or graviton particles. Proofs of these in our science have simply been misinterpreted, as explained in Mark's book.

Mark McCutcheon does a great job of debunking Standard Theory in his book *The Final Theory. Links to the debunk of our actual science* can be found at the end of chapter 1. It has been written for laypeople, for anyone whether they have a scientific background or not. It re-writes our entire physics and is the only working theory of everything that truly has an answer for everything in our science.

Dismissing Atomic Expansion Theory without reading the book, could cost humanity a great deal. And if you reject everything else from my book *New Age Physics,* at the very least you should give *The Final Theory* a fair chance. Our current models have served us well, they were describing our reality well to a certain extent, but with Atomic Expansion Theory we could achieve much more.

3.13 Two best proofs of Atomic Expansion Theory - The levitating slinky and ball

It is very difficult to identify a proof for a theory in physics, when other theories that might be incorrect, or misinterpretations of physical phenomena, can still model reality very well, and equally explain what we see. Atomic Expansion Theory can explain all the so-called proofs of Einstein's theories, and can re-interpret everything stated within Quantum Mechanics, from the point of view of expanding and contracting matter.

They are all theories after all, supposed to model perfectly all identified phenomena in nature. And you can always tweak these theories to infinity, so they still model what we see. For example, by inventing wild and unlikely concepts such as dark matter, dark energy, gravitational waves, graviton particles, black holes, singularities, etc. Better that, than admitting that a theory is incorrect, I suppose. But when exactly do we draw the line and admit that the theory is wrong?

The key to prove Atomic Expansion Theory, is to find proofs where the expansion of the planet and all objects, explain something that no other theory can explain. So it should be a consequence of the expansion of matter, and show what is happening behind the scenes, to justify what we see in our resulting reality.

I came up with the two best proofs for Atomic Expansion Theory that I could think of. I do not guarantee that in time, people would not use other theories to justify what is happening, and state that these are not proofs after all. It would not disprove Atomic Expansion Theory, it would simply mean that other theories can also explain them. However, they better explain them convincingly, without resorting to magical thinking. I'm pretty sure levitation is not part of Standard Physics, no matter how someone somehow could justify it.

3.13a The physics of a stretched suspended slinky being dropped

No proof appears to be convincing enough, it seems. Which brings us to the physics of the drop of a stretched suspended slinky. Good luck explaining convincingly that particular phenomenon, using either Newton or Einstein's theories of gravity.

First you need to watch on YouTube these four videos from *Veritasium,* showing in slow motion what is happening when you drop a fully stretched slinky from the top of a building:

Does a Falling Slinky Defy Gravity?
https://youtu.be/uiyMuHuCFo4
Supersized Slow-Mo Slinky Drop
https://youtu.be/JsytnJ_pSf8
Slinky Drop Answer
https://youtu.be/eCMmmEEyOOO
Slinky Drop Extended
https://youtu.be/oKb2tCtpvNU

When you drop an extended slinky from high up, film it and play the video in slow motion, the base of the slinky does not move at all, until the top reaches the base. Almost as if it was free floating in space, as in Einstein and Atomic Expansion Theory, which are geometry based theories, with the exception that it is actually levitating for a while.

This experiment is even visible to the naked eye, from a drop of your own height. There's no need to drop the slinky from the top of a building, or to film it first in order to slow it down. You will see, the bottom levitates, it stays there suspended in the air, until the slinky is fully contracted back, and only then, will it fall down.

The first obvious issue is that levitation is impossible in Newton and Einstein. As soon as an object is dropped, the entire apparatus must immediately start falling down. This is true whether gravity is a force acting at a distance, or is explained by geometry, where any dropped object must immediately follow the curvature of spacetime.

You would not expect a bearing ball dropped at the top edge of a funnel, or of a garden trampoline, not to move for a while, before starting to fall towards the middle. Slinkies and springs don't have a special status when it comes to gravity, simply because when they are stretched then dropped, they not only fall, but they also contract back to normal. And that is key.

The second obvious issue is that, it is impossible for a stretched slinky or a spring, suddenly freed from both ends, not to have both ends come back together. One end cannot suddenly remain immobile in the air, while the other end comes back towards it. This is a serious misinterpretation of what is happening here.

In Einstein, the bottom end of the slinky is not attached to anything. And granted that there is no force acting at a distance upon it, and it is free floating, but once dropped from the top, it should still immediately follow the curvature of spacetime to its destination, which is the ground.

These two issues are common sense. The first issue is telling us that neither Newton nor Einstein can explain this observation, which is levitation, hence they cannot be correct explanations for gravity. The second issue is telling us that, despite what we observe, the bottom end of the slinky must be going up, as the top end is coming down, as the slinky contracts back together while falling.

It cannot possibly matter, if the slinky is more stretched nearer the top, and more contracted towards the bottom, with a centre of mass nearer the bottom. Both ends must always come back together, even if they don't do so in equal measures or at the same rate.

Fine, I hear you say. The slinky is not falling immediately, because the bottom end is going up, as the slinky contracts back, which explains why it remains suspended in the air for a while. However, this is still magical levitation.

No matter what anyone might say, gravity does not wait for information to come from the top of the slinky to reach the bottom, before starting to fall, no matter where its centre of mass is. This is a desperate and illogical argument to explain the unexplainable, in the absence of a valid theory in physics that could explain what is really going on.

And the bottom end cannot remain immobile floating in the air, even if it is going back up. The entire slinky must immediately start to fall down under both current theories of gravity, even while it is contracting back to normal.

In the *Supersized Slow-Mo Slinky Drop* video, they have an experiment meant to reflect the other, but without any gravity involved. They put an unstretched spring horizontally on a table, hit one end, and we can see that the other end remains immobile until the spring is fully contracted, then it bounces back. This is an entirely different experiment not equivalent in any way to the first one.

First, there is no real stretching of the spring whatsoever. So, you would not expect both ends of a spring to come back together, when you let go of it, when the spring is not stretched.

Second, not only they have eliminated the downward force of gravity, but they also removed the other upward force against gravity, while the object is being held. There are two forces here acting on the slinky. So, whether one end of the spring remains immobile or not while horizontally, in that particular experiment, is hardly equivalent to what would happen with gravity, a force holding the slinky at the top, and stretching.

The linchpin of the entire phenomenon, is the fact that not only the slinky falls down due to gravity, but also that it contracts back at the same time. If you eliminate the stretching as well as gravity and the holding force, then you are comparing two different phenomena.

A more accurate experiment would have been a horizontal spring tied at one end, being dragged on the table by that end until it is stretched, and then let go. Then both ends would have come back together during the contraction of the spring, no end would have remained immobile.

And this is precisely what is happening that we cannot see, when the slinky is suspended from a rooftop, because it is only happening behind the scenes, where we can see the expansion of the Earth.

So here is the solution that no other theory, I feel, could explain. Once the slinky is dropped, the base must be going up while the top is going down, while the entire slinky contracts. But the bottom end must be going up at the exact same rate as the expansion of the Earth. Which is not surprising, since the stretching corresponds precisely to the measure of gravity on Earth.

This is because, the force holding the slinky at the top before the drop, is dependent on the expansion rate of the planet pushing it upwards in the first place. The planet is constantly expanding at a rate of 4.9 metres per second, pushing everything upwards, and this is what causes the initial stretching of the slinky while it is being held.

Consequently, it seems like the bottom end of the slinky is not moving at all, when in fact the base is going up. The important difference here, is that the slinky is free floating in the air as soon as it is dropped. It is not being attracted to the Earth, and it is not following the curvature of spacetime. It is the Earth that catches up with the entire apparatus in its expansion, once the slinky is no longer being pulled upwards while suspended.

And since the base is going up at the exact same rate as the expansion of the planet, it appears as if the slinky is levitating in the air. The base of the slinky is not going up from our point of view, but it actually is behind the scenes, if you could see the expansion.

It would be interesting to try a variant of the original experiment to see the result. For example, if someone on the sixth floor holds the top of a supersized slinky, normally it would stretch all the way to the second floor. But if someone on the fourth floor brings up the bottom half of the slinky, holds the base there, and then both persons let go of the slinky at the same time, then what would happen to the base of the slinky?

In the original experiment, the base of the slinky appeared immobile and levitating, because in the background, it was going back up at the exact same rate as the Earth's expansion. In the second experiment, the stretching of the slinky no longer corresponds to the Earth's gravity, since the base is held at the mid-point. It is therefore possible that the slinky would start to fall down immediately. However, it would do so slower than would be expected from normal gravity, since the base would still be going upwards, while the slinky contracts back to normal.

This said, since the stretch is now less pronounced than before, and both ends would be coming back with less force, it might compensate for the difference in the two experiments. After all, the stretch still corresponds to the force of gravity, it is just that the slinky is now held at two different points. As a result, the base might still levitate and remain immobile in the air, until the top reaches the base, and then fall down. If you attempt it, please let me know the result, so this point can be settled.

And now, let's move on with another prediction that could doubly prove Atomic Expansion Theory, when we replace the slinky with a ball. Which brings us to our second best proof.

3.13b The physics of a suspended ball being dropped

If a person on the top of a building is holding a ball, then obviously the building, the person and the ball, would be going up at a rate of 4.9 metres per second, being pushed upwards by the expansion of the Earth.

So when the person drops the ball, and the distance between the ball and the ground reduces in an acceleration of 9.8 metres per second square, the expansion of the Earth would first need to overcome that initial speed of the ball going upwards.

And while it is doing so, just like with the slinky before, the ball should appear immobile in the air, as if levitating, in the first half second it will take for gravity, to overcome that initial upward speed of 4.9 metres per second.

This is completely different from Newton or Einstein, where the dropped ball should immediately start to fall back to Earth. Because in their theories, there is no such thing as an expanding Earth justifying gravity, and the ball held in the person's hand would not already be going upwards at a rate of 4.9 metres per second, before being dropped, since this is a consequence of the expansion of the planet. So if we could witness this, it would be the perfect proof for Atomic Expansion Theory.

I don't know if from these limited heights, we would be able to prove a levitating ball, just before it starts to fall back down. However, the levitating act of the slinky is so striking and obvious, it gives me hope that we might also be able to witness and measure the levitating act of the ball. A speed of 4.9 metres per second upwards before being dropped, as could be seen behind the scenes, is quite a measurable quantity, so maybe this could be tested and proven.

Now, there is another issue. Measuring this using our actual equations for gravity, or even the Atomic Expansion Equation, might not be sufficient. Because these equations might have been derived from empirical evidence, from how long a dropped object takes to reach the ground, through simple observation. And without these equations having been established, while being aware of that initial upward speed of any held object, these equations might already be measuring it by default.

For example, Mark McCutcheon has developed his Atomic Expansion Equation, based on empirical evidence of how long it takes an object to fall back to Earth. Most probably Newton and Einstein did the same, since their equations needed to match observation. Even the expansion rate of the atom, and of everything else, is based on how fast an object falls back to Earth, and might need to be re-adjusted slightly in light what I am saying now.

Even an object falling from a plane has an upward speed, as planes must constantly be going up in order to counteract the expansion of the planet. Even an object dropped from the International Space Station must be going upward, because behind the scenes, all orbits are constantly enlarging away from Earth, while all objects are being accelerated as well. Anyway, an object being dropped from the space station, would simply continue on the same orbit as the station, since it would already be going at the right speed for that altitude, to continue on the same orbit.

Therefore, any kind of measurement comparisons for this, should be done in space, far from the planet. So we could compare if our equations bring similar results, for the gravity of two objects totally free in space, and the gravity between the ground and a held ball, before being dropped, while standing on an expanding planet.

This said, the levitating act of the ball in slow motion, might be as obvious as the slinky one, for a long half a second, before the ball starts to fall. And that prediction and fact alone would be all the proof we need. No calculations using equations are required to prove it.

3.14 Atomic Expansion Theory concepts and orbit simulations on YouTube

It might be helpful to check out these orbit simulations and discussions about the concepts of Atomic Expansion Theory on YouTube. Mark McCutcheon and I are grateful to these talented artists, who were intrigued enough to put their mind into making these simulations and videos, and to the thinkers who are further developing our theories:

3.14a Cruz deWilde

- www.avantgravity.com
- www.youtube.com/c/cdewilde
- Turning Gravity Inside Out
- Gravity as a Result of Expansion
- Gravitational clustering as a result of particle expansion

3.14b Life, Everything And The Universe

- www.youtube.com/channel/UC5835CWoYWpQrv05ahKbugQ
- Think Gravity is a force or warped spacetime? This realtime simulation might convince you otherwise.
- Slow motion footage reveals one of the paradoxes in theories of gravity. See it with your own eyes.
- Powerful Gravity Thought Experiment. You might see things differently after this. (Inflating Balloon)

- A Proof, Using 3 Logic Steps. That Current Gravity Theories Are Indeed Invalid
- Visual confirmation that gravity has absolutely ZERO force
- How Gravity Might Really Work. All The Paradoxes Explained From An "Outside The System" Perspective.
- Quick Push Pull Gravity Animation

3.14c The late Gerald Clark's series about gravity featuring an interview with Roland Michel Tremblay

- www.youtube.com/channel/UCFUH_0JRBPG9g5k5M3AgJ4g
- FREE Gravity Series Episode 1 [Gerald Clark Premier Content]
- FREE Gravity Series S1E3 "The Final Theory" Part 1
- Gravity Series S1E3 Final Theory Explained Supplement
- Gravity Series Final Theory of Expanding Matter and Orbits
- Atomic Expansion Theory - A Theory of Everything - Full Interview Gerald Clark and Roland Michel Tremblay - https://youtu.be/8lj1F8UqPc8

3.14d Gerald Clark's Premium Content requiring subscription, except for the free ones indicated, which are also on YouTube

- www.geraldclark77.vhx.tv/gravity
- Ep1: Gravity Series Episode 1: What is Gravity? (Free)
- Ep2: Gravity Series Episode 2: Gravity Pioneers
- Ep3: Gravity Series Episode 3: Final Theory Part 1 (Free, part of my interview with him)
- Ep4: Gravity Series Episode 3: Final Theory Part 3
- Ep5: Gravity Series Episode 3: Final Theory Part 2

- Ep6: Gravity Series Episode 4: Search for Gravity Waves
- Ep7: Gravity Series Episode 5: Gravity and You (Free)
- Ep8: Solar System Helical Orbit and Satellites

3.14e Chris Freely (The Cosmic Fool)

- www.chrisfreely.com or www.chrisfreely.blogspot.com
- www.youtube.com/channel/UCE1h_z9F-fyLIX-IOmRKcjQ
- Expansion Theory Science The Unified Field Theory

3.14f Ian Moore (Ianto)

- www.candigram.co.uk
- www.youtube.com/channel/UCPQR8jv6Xk3kpOsBm4A_iHQ
- Proof of NoThing
- Proof of SomeThing

3.15 Standard Theory and Atomic Expansion Theory Maps

The next two figures are full page maps with larger versions available online:

Figure 15 Standard Theory map
www.themarginal.com/map1.jpg
Figure 16 Atomic Expansion Theory map
www.themarginal.com/map2.jpg

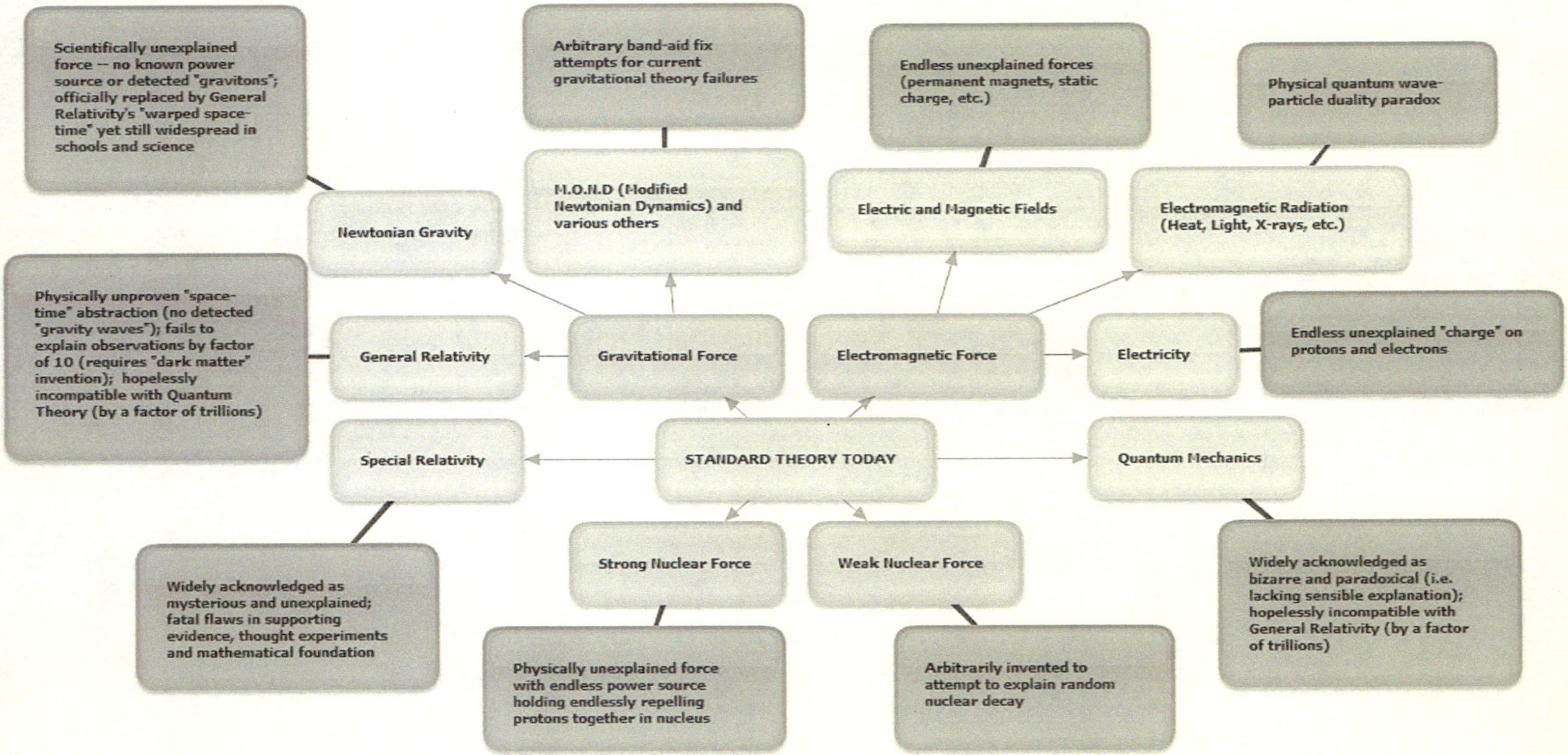
STANDARD THEORY TODAY
Gravitational Force
Electromagnetic Force
Strong Nuclear Force
Weak Nuclear Force
Special Relativity
Quantum Mechanics
Newtonian Gravity
M.O.N.D (Modified Newtonian Dynamics) and various others
General Relativity
Electric and Magnetic Fields
Electromagnetic Radiation (Heat, Light, X-rays, etc.)
Electricity
Scientifically unexplained force -- no known power source or detected "gravitons"; officially replaced by General Relativity's "warped space-time" yet still widespread in schools and science
Arbitrary band-aid fix attempts for current gravitational theory failures
Endless unexplained forces (permanent magnets, static charge, etc.)
Physical quantum wave-particle duality paradox
Physically unproven "space-time" abstraction (no detected "gravity waves"); fails to explain observations by factor of 10 (requires "dark matter" invention); hopelessly incompatible with Quantum Theory (by a factor of trillions)
Endless unexplained "charge" on protons and electrons
Widely acknowledged as mysterious and unexplained; fatal flaws in supporting evidence, thought experiments and mathematical foundation
Physically unexplained force with endless power source holding endlessly repelling protons together in nucleus
Arbitrarily invented to attempt to explain random nuclear decay
Widely acknowledged as bizarre and paradoxical (i.e. lacking sensible explanation); hopelessly incompatible with General Relativity (by a factor of trillions)

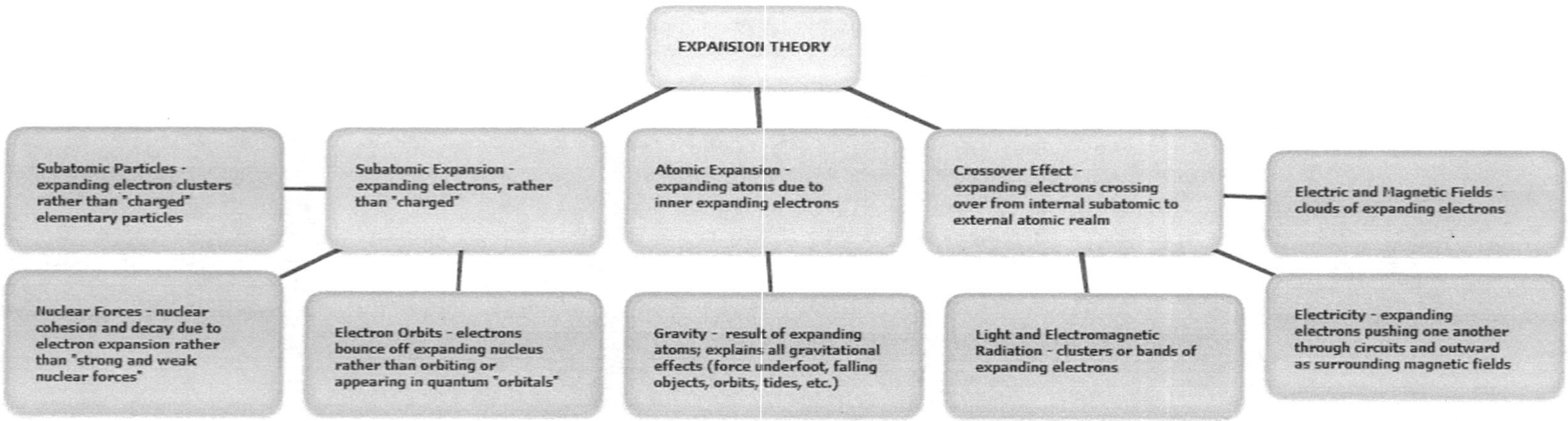
EXPANSION THEORY
Subatomic Expansion - expanding electrons, rather than "charged"
Atomic Expansion - expanding atoms due to inner expanding electrons
Crossover Effect - expanding electrons crossing over from internal subatomic to external atomic realm
Subatomic Particles - expanding electron clusters rather than "charged" elementary particles
Nuclear Forces - nuclear cohesion and decay due to electron expansion rather than "strong and weak nuclear forces"
Electron Orbits - electrons bounce off expanding nucleus rather than orbiting or appearing in quantum "orbitals"
Gravity - result of expanding atoms; explains all gravitational effects (force underfoot, falling objects, orbits, tides, etc.)
Light and Electromagnetic Radiation - clusters or bands of expanding electrons
Electric and Magnetic Fields - clouds of expanding electrons
Electricity - expanding electrons pushing one another through circuits and outward as surrounding magnetic fields

FOUR

New models for atoms as galaxies, electrons as solar systems, super-neutrinos as stars, neutrinos as planets, sub-neutrinos as moons, and sub-sub-neutrinos as comets, asteroids and rocks

4.1 Several scales of universes composing each other, ruled by the exact same laws of physics

For Mark McCutcheon and Standard Physics, there are only two scale universes. The point of theoretical physics is to find a Theory of Everything that would unite the laws of physics of the very small, ruled by Quantum Mechanics, with the laws of the very large, ruled by the Theories of Relativity.

Anyone reading the book *The Final Theory* of Mark McCutcheon, will agree that he succeeded in unifying these laws of physics, so that the same laws can be applied to both scales. For a start, he eliminated the need for Quantum Mechanics, Einstein and Newton from our physics, and everything is explained by the expansion and the motion of electrons, atoms, molecules, and the larger objects they compose, such as planets and suns.

However, for me something is still missing, and we can go even further in our attempt to truly have the exact same laws of physics working at all scales. I never believed that the electron and the neutrino were the smallest particles ever, that they were some sort of point particles without size, mass or volume. I just looked at what we saw in a microscope, and what we saw in a telescope, and it was obvious to me that it must all be the same thing at two different scales. I think I was born with this idea, like many before me, and it simply won't go away.

How could the very same laws of physics equally applied at both scales, not produce the very same patterns at all scales? If one scale is radically different from the other, like electrons are jumping on the nuclei of atoms instead of orbiting, or the atom is a foreign dimension to our reality, then there is no hope for a theory of everything, we would need different laws for different scales. It would be strange indeed, if the smaller scale universe was not the same as the larger scale universe, especially when physics is now defined by the expansion of matter.

And for me there were always more scales. The very small scale universe must be composed of an identical universe at yet another smaller scale. And the very large must be composing an even larger scale universe, identical to what we are looking at in the night sky. Just like fractals, the same composing the same, at an infinity of scale universes.

I don't think we will ever know if there is an infinite amount of scale universes out there, composing more of the same. There could be a prime scale universe, with indivisible point particles, composing the very first scale universe in the infinitely small. And maybe there is a final large scale universe, where the patterns break down and more of the same is no longer being formed.

Nevertheless, in this chapter I will explain exactly how the very small is identical to the very large, how the very same laws of physics Mark McCutcheon has established, are to be applied exactly the same way at both scales - including gravity and the natural orbit effect of the geometry of expansion - and how the very large is almost certainly composing an even larger universe at a higher scale.

And I am making three great predictions that will prove this if ever discovered. First, we will find planets orbiting between solar systems, bonding them together. Second, we will find solar systems orbiting between galaxies, bonding them together. And third, we will find solar systems orbiting between clusters of galaxies, bonding these clusters together. I hope you can find a powerful telescope and start looking for these rogue solar systems and planets, and hopefully we will be able to establish their trajectories in space.

It is possible that we will find that clusters of galaxies are linked together by rogue galaxies orbiting between them, instead of rogue solar systems or stars, but it is unlikely. For this to be true, individual molecules would need to be bonded together by atoms orbiting between them. As far as I know, molecules are more likely to be bonded together through an exchange of electrons orbiting between them, instead of atoms. But who knows exactly what is happening between individual molecules, and how they remain bonded together, or repel each other?

My best guess is through an exchange of electrons, but it could be through an exchange of orbiting atoms between them instead. Therefore, we should keep an eye out for a fourth possible prediction: orbiting galaxies between clusters of galaxies, bonding these clusters together. That possibility should be easy to confirm or reject. But, as I mentioned, this is not likely to be the case, even if it would seem more logical.

4.2 Is the very small (sub-neutrinos, neutrinos, super-neutrinos, electrons, atoms) similar to what we see in the very large (moons, planets, suns, solar systems, galaxies)?

Mark McCutcheon and Standard Physics don't see how the very small scale could be compared or be the same as the very large scale universe. They cannot find much in common between neutrinos and planets, electrons and solar systems, atoms and galaxies, and molecules and clusters of interconnected galaxies. But this is the key to what was missing from our science, how the very small scale is identical to the very large scale, and now the same laws of physics can truly be applied to both scales.

Atoms appear to be robust objects that can be squeezed and can bounce, and can even push back from within their subatomic realm, as a way to explain inertia. If you squeeze several atoms together, they will attempt to get back to their normal shape of being un-squeezed, and will push back vigorously. For McCutcheon, electrons are bouncing on top of the nuclei of atoms, and bouncing between the nuclei of different atoms in chemical bonds. If you squeeze atoms, the electrons bounce harder and faster. Atoms and electrons are as solid as they come.

Solar systems and galaxies look rather fragile, presumably they could never achieve such a feat. If two solar systems were to meet in space, or two galaxies, there would be no bouncing back involved, they would all come crashing down into each other. I will now demonstrate that this does not have to be the case, and that galaxies and solar systems can be as strong as atoms and electrons. This is achieved through the exchange of solar systems orbiting between galaxies, and the exchange of planets orbiting between solar systems, which can keep them safely apart and even push back. Of course, it does not always work, systems can be unstable, and sometimes we will see two galaxies or two solar systems crashing into each other.

There are at least two different scales of universes from what we can see in a microscope and a telescope, and no matter any relativistic motion that could shrink or enlarge you from the point of view of others, there will still be different scales and sizes of universes. A small universe is composing a larger one, which in turn seems to compose an even larger one, all this within a world of expansion, expanding at the same constant rate at all scales.

Possibly atoms are galaxies, and maybe electrons could actually be solar systems. The suns, moons and planets then would simply be neutrinos, and together they would form the nucleus of the electron. Most solar systems have two stars instead of one, and some have up to seven stars. And planets can become stars, if they become massive enough. So, they would all be considered neutrinos. The electron's nucleus would be made entirely of neutrinos, just like the atom's nucleus is entirely made of electrons. In Atomic Expansion Theory, protons and neutrons within the atoms are made of electrons, which means that we live in an electronic world, just like in a computer simulation.

I even have a psychic friend in Los Angeles, who had a vision that the entire universe is in fact a biological cell at a larger scale. Who knows? You will see that the idea is not as crazy as it seems at first sight.

4.3 Comparing subatomic and atomic particles to what they compose at the larger scale

In the first type of expansion of matter, atoms and electrons are constantly expanding independently from the other four types of expansion and contraction of matter, this is defined as their inner expansion. More specifically, they expand independently from relative motion, the fifth type of expansion and contraction of matter, which will be covered in subsequent chapters.

This is how we can have stable scale universes, like one made of electrons and atoms, composing another made of solar systems and galaxies. And since the inner expansion rate of stars and planets, depends on the inner expansion rate of what composes them at the smaller scale, then everything in the universe must expand at the very same Universal Atomic Expansion Rate, no matter the scale.

Those scale universes, no matter if we can view them smaller or larger through the relativity of motion, remain overall proportionally at their normal size, while still expanding at a constant rate. I mean, they don't ever reduce in size, except through relative motion, and that is simply an optical illusion, which depends on the observer's frame of reference.

Relative motion is the expansion or the contraction of objects, as they move towards us or away from us, instead of covering any distance. Is it then plausible, that if something like a spacecraft shrinks enough, through simple motion, that it might eventually reach a size smaller than one electron? It sounds like a stretch, that trillions of atoms composing such a spacecraft, could shrink back to the size of one electron, through relative motion alone. And don't forget, it would only be shrunk from the point of view of someone else, somewhere else.

Objects can also grow or shrink in size in the third type of expansion of matter, where electrons, atoms and molecules all expand or contract. But they never do so that much, and then they simply move to another dimension, practically a different reality. Electrons and subatomic objects can equally grow or shrink in the second type of expansion, but there again, never that much. And we're not talking about objects here, since this growth and shrinkage of electrons and subatomic particles, happens outside the atoms and molecules, leaving them unaffected.

I have not discussed the fourth type of expansion of matter much, so far. This is when we open our eyes, and we psychologically create the physical universe within our minds, and project it out there in a quasi-instant expansion, where all our senses can perceive it as a physical reality. Just like when we turn on a computer, and a simulated world or virtual reality is being loaded and projected onto the screen. I will discuss this further later on.

To shrink an object to the size of an electron in real time, or to expand it to the size of a cluster of galaxies, we would need to be able to manipulate the Universal Atomic Expansion Rate of matter, the first type of expansion. I have discussed that topic in subsection *3.11a The first perspective - The God's viewpoint with expansion - And can we change the expansion rate of matter?*

Objects radically do change in size because of the relativity of motion. At that point there are an infinite number of different sizes in the universe, as many as there are points of view from which we can perceive the universe. For now, I am only concerned with the different sizes or scales of universes, ignoring the optical illusion of relative motion, which is, as mentioned earlier, the fifth type of expansion and contraction of matter.

If you can visualise a starting point to all realities, as Mark McCutcheon suggests in *The Final Theory,* as simply the sudden appearance of a lot of electrons in the universe, distributed randomly across a certain space, then the geometry of expansion of these expanding electrons would go on to form atoms, molecules, then beings and objects. Objects would be any of our everyday objects, like a table or a TV, but also planets, stars, solar systems and galaxies. Of course, it may be neutrinos that appeared randomly out of nowhere one day, instead of electrons, in this creation type event of the universe.

At this point, we find ourselves seeing at the larger scale, what appears to be very similar to what we observe at the smaller scale, and yet it is all interwoven together, since at the base of it all are the electrons and atoms. Those larger objects, such as solar systems and galaxies, which now reflect the smaller ones composing them, could then be considered another scale or size of universe.

It is conceivable that solar systems and galaxies also form molecules, beings and objects at a higher scale, or at another size of universe, and that would be yet another scale universe. A reality made out of larger particles, which are in fact solar systems and galaxies, populated by beings composed of solar systems and galaxies, just like we are composed of electrons and atoms.

We could conjecture that the way electrons bond together to form particles, like protons and neutrons, or how they form clusters in light and heat, could be similar to the way atoms bond together to form molecules. Orbiting neutrinos within the electrons would orbit between the nuclei of two or more electrons, attracting or repelling other electrons, depending on the number of neutrinos in their higher orbit, in order to reach a balance through the geometry of expansion. This would be the same way electrons within atoms are orbiting from one nucleus to another, to get atoms to form chemical bonds.

Possibly then, electrons have a similar structure as the one of the atoms. A nucleus with orbiting particles, some of which orbit very far from the nucleus, which could get caught in the expansion of other particles nearby, following the natural orbit effect of the geometry of expansion. If there is a need for balance, these smaller particles in higher orbits will then orbit between both particles, bonding them together, in order to form larger stable particles.

Standard Physics appears uncertain about the existence of neutrinos, many believe in their existence, and now they even have a mass. But for Standard Physics, electrons would not be made of neutrinos. Electrons are still indivisible, unless they are annihilated and vanish into a burst of energy. Which I surmise must simply be, that smashing an electron, ejects at once all its neutrinos out into space, like atoms do with their electrons in a nuclear explosion. Remember that in Atomic Expansion Theory, everything is made of electrons, including the protons and neutrons within the atom. While for me everything is instead made of neutrinos, including the nucleus of an electron.

Mark McCutcheon is still unconvinced about the existence of neutrinos, at least until proven without any doubt. For him, there is nothing smaller than an electron, and so, electrons are indivisible, and they don't have any kind of real substance in our reality. Electrons don't annihilate in a puff of energy either, they just take the form of radiant energy, which is electrons in clusters, freed from the subatomic world through the crossover effect, which is the second type of expansion and contraction of matter. For Mark, it is more likely that electrons will be found in groups wherever we look, instead of single electrons travelling alone in space.

We cannot tell if like fractals, there could be an unlimited number of scale universes composing each other, or if there are only a limited number of them, or if the very small and the very large are all that exist, as Standard Theory and Mark McCutcheon tell us. If electrons and atoms form stars and galaxies, why could stars and galaxies not form something at a higher scale?

Are there truly excellent arguments against the possibility that there could be something smaller composing electrons, and even neutrinos? I haven't found these arguments if they exist, apart from the fact that, at that even smaller scale, we might never be able to see or detect any such sub-subatomic particles, as we are struggling to detect neutrinos as it is.

In my opinion there are neutrinos composing electrons, because I feel something smaller still, within the electrons, must be able to attach and detach from other electrons. Or else, electrons would not have what Standard Physics calls a charge, or a way to either attract or repel other particles.

In Atomic Expansion Theory there are no charged particles. There are only particles that attract or repel each other, through a balance or imbalance of electrons, orbiting between the atoms' nuclei. And outside the atom, we have clouds of electrons of different densities, either pushing against each other, if they are the same size, or attracted to each other, if they differ in size. And now, possibly, we have neutrinos orbiting around and between the electrons' nuclei, and there may be clouds of neutrinos outside the electrons, like there are clouds of electrons outside the atoms.

It might be tempting to believe that these clouds of neutrinos might even form very weak magnetic fields, the way electron clouds form our more familiar magnetic fields. However, neutrinos, like planets, are neutral particles. They don't have smaller particles orbiting between them and other objects, capable of pushing back or attracting other particles and objects. The reason electrons can form magnetic fields, either pushing back or attracting other particles and objects, is because of the exchange of neutrinos between them, which explains how electrons and other particles can have a positive or negative charge.

This is an extremely important point. How could electrons have a negative charge, or a positive charge as positrons, if somehow they are just balls or point particles jumping around? Would they not then be neutral particles? How do they attract or repel particles or each other, or attach and detach from particles or each other, or form particles? Through magnetic fields? Electrons are magnetic fields, they form them in clouds of electrons, so that could not be how.

McCutcheon would say they have no charge, and don't attract or repel particles or each other. They just bounce from one nucleus of an atom to another, or simply push against each other on wires, or in space in clouds or clusters. They move while seeking balance in any system, from where there is a surplus to where there is a depletion, equalising in size and numbers when they come into contact. However, how do electrons remain in clusters in light, while these clusters are pushing each other? How do they remain attached to form protons, neutrons, clusters and clouds, if they have no charge?

It is possible that a bunch of electrons coming together in a certain number, would form into a stable proton, neutron, or cluster in light, but I find that hard to believe. It would be like assembling together many thousands of planets, and hoping that the force of expansion would actually give us nice stable particles of the same size, all glued together, pushing each other while keeping their stability and individuality.

Would they not instead turn into one big mass of planets, one gigantic planet? How do they keep stable in separate larger particles, all of the same size, all lined up against each other? In molecular structures such as crystals, all the atoms are interconnected and kept apart through the exchange of electrons between them. It has to be same for electrons, they must be exchanging smaller particles.

I believe electrons are smaller versions of atoms. Electrons have a nucleus made of neutrinos, they have orbiting neutrinos, they have neutrinos orbiting from one nucleus to another, and in this way they can create bonds. In this way they can have an electric charge, meaning they can attract or repel other particles through an exchange of smaller particles. And that is my new model of the electron.

In Standard Physics, we don't really have a model for the electron. They are so small, they have never been seen. We cannot even be sure they exist, apart from measuring levels of energy, as they interact with other particles. They are considered shapeless, with no mass or volume, small balls, point particles, or made of tiny vibrating strings. Take your pick.

And now, the electron has a new model, it is like a smaller but simpler version of an atom, where orbiting electrons are replaced by orbiting neutrinos. I also conjecture that there must be something even smaller than sub-sub-neutrinos, but not at the same scale. Sub-sub-neutrinos, such as comets, asteroids and rocks, would be made of particles at another much smaller scale. And all these particles are simply constantly expanding at the exact same constant rate, no matter the scale.

I believe that all these particles, and larger objects in the very large similar to them, are more or less of the same structure, even the spiralling shape of a galaxy when viewed from here. This structure is the same as the one of a solar system, even if it is more elaborate, because there is much more matter, orbiting much slower.

Galaxies, atoms, solar systems and electrons would all be of that similar structure. They must all have a sort of nucleus made of a concentration of one particular particle, which at the higher scale looks like a sun, with orbiting similar particles that can get them to attach and detach from similar objects, through orbiting between two or several nuclei of these similar objects. And through the geometry of expansion, gravity and the natural orbit effect, no matter the scale, they would form bonds like molecules, and eventually form larger objects at a higher scale, or if you wish look-alike particles at a higher scale.

We get some idea of how much stuff can go within one such particular object, such as a galaxy. Our solar system does not appear to have as much clutter as a galaxy, but maybe atoms do, and we just don't see it or realise it. Solar systems would be more akin to electrons, while galaxies to atoms.

At any rate, they all appear to have a nucleus with orbiting smaller particles around them, while the nucleus is made of the same stuff as the orbiting particles. Which would make sense, if the force of expansion of all matter, eventually gets a lot of the smaller bits of matter to stick into larger objects in the nucleus. The structure is generally flat, but in its revolutions and stages of formation, it might appear or even sometimes be round, oval or spherical.

Consequently, atoms must be mostly flat objects, just like most galaxies are. Atoms only appear spherical to us due to their high speed, viewed from here. Electrons must equally be flat objects, just like our solar system is. Indeed, all the orbiting planets orbit on the very same horizontal plane, called the invariable plane. Pluto was the only planet on a strange orbit, radically different from the others, but it is no longer a planet. Because of Pluto's out of alignment orbit, it was thought it came from outside the solar system, and was not part of the original solar system formation. Maybe we'll find interesting things on Pluto.

To confirm that atoms and electrons are flat, we would first need to be able to see at that small scale, and if we have not succeeded by now, will we ever? However, maybe there is no need to see at that small scale, when we can look up at the night sky, and see the very same thing at a larger scale, and in slow motion as a bonus. Then again, we're not exactly expert at spotting rogue planets outside solar systems in the universe, the issue is that they don't emit light, therefore they are nearly impossible to find. Rogue solar systems outside visible galaxies, should be far easier to spot.

In the case of galaxies, viewed from here, we generally see flat spiralling arms full of solar systems, instead of precise orbits like our planets around the solar system. However, this does not change the fact that solar systems within a galaxy, taken individually on their own, are simply orbiting around the galaxy centre, just like the planets do around the Sun.

They just happen to be in arms formation, because of the way a galaxy forms. A build-up of pressure at the centre of a massive object, will eject the matter at both poles in two long jets of matter, which then become the galaxy arms. The straight arms then start orbiting around the galaxy centre, while the matter nearer the centre orbits faster than the matter at the edge, leading to spiralling arms. I explained galaxy formations in section *3.4 Gravity and the formation of galaxies.*

Galaxies of different shapes, might simply be in an early stage of formation, or might be the result of a recent collision with another galaxy. Or they recently collapsed on themselves, and through pressure have released two jets of matter at both poles, which now we see as two bars, while the arms are sort of starting to form. Perhaps atoms can be of all these different shapes and structures as well. Possibly then, electrons, atoms, solar systems and galaxies are all of a similar structure, the one of a nucleus with orbiting particles.

So how many scales of the same structure are we seeing from here, or could potentially see, or at least that we can surmise exist? What constitutes a different scale of universe? It seems logical to say that neutrinos, electrons, atoms and molecules are one scale, while planets, solar systems, galaxies and galactic molecules at a larger scale are another.

Any smaller or larger scale would be particles radically smaller or radically larger, to an extent that our technology could never even hope to measure or observe. And yet, let's assume that these even smaller and larger scales actually exist. At a minimum, so far, we can at least surmise the existence of four scale universes.

At the very least, Seth in Jane Roberts' books, certainly hinted that what we see in the night sky, is forming another being at a larger scale. Maybe he was referring to our definition of God. It certainly would explain why God would be everywhere in every particle that exists, and would be All That Is.

I do wonder sometimes if there are civilisations very much like us, composed of particles from two scale universes down, in the very very small, living on particles within us. We better not die too quickly then. But they are vibrating so fast compared to us, their time rate must be much faster than ours. They could probably live the equivalent of a few billion years within our own lifetime. Plenty of time then for these civilisations to be born and go extinct, before we die and go extinct ourselves.

4.4 A galaxy is the same as an atom

With this new possible vision that everything that we see through a microscope might be of a similar structure than what we see with a telescope, next is my new way of comparing what we see at the smaller scale with what looks similar at the higher scale.

A galaxy is made out of solar systems, just like an atom is made out of electrons. You might believe there are not as many electrons within an atom as there are solar systems in a galaxy, however, we have never seen what the inside of an atom looks like.

Our model of the atom has changed several times in the last two centuries alone, it has even changed again in the last 20 years, with Mark McCutcheon's new model of jumping electrons over the nucleus. And even the supposed protons and neutrons composing the atom are made of electrons, according to Atomic Expansion Theory.

In the Periodic Table of Elements, when describing how many protons, neutrons and electrons an atom contains, we may in fact be talking about certain quantities of packets of electrons. One proton or one neutron could refer to a certain number of electrons bonded together, they don't necessarily need to be of a defined shape, such as a sphere or a point particle.

Just like when we say that just one electron is being exchanged between atoms. It only means a certain quantity of energy being exchanged, that we can measure with our precise instruments, since we have never seen what an electron looks like. When suddenly the energy level switches and doubles, then we say there are two electrons. Is it really the case, though? Who is to say if it is not a whole bunch of electrons bonded together, that correspond to this energy level?

I encountered the same problem studying the neutrino. They often talk of one neutrino being ejected in the beta decay of neutrons, to account for the energy being released, along with one proton and one electron. However, with variable amounts of energy being released from one decay to another, it seems clear to me that we are not talking about one neutrino being ejected, but several instead. Or, these neutrinos are of variable sizes. And how could a decaying particle eject neutrinos, if that particle was not made of neutrinos in the first place?

We only ever guessed anyway such quantities, because what we measure is instead amounts of energy being gained or lost, as molecules are formed, as chemical reactions happen, or particles decay. We have never seen or tracked down an electron or any of these particles, we only measure energy input and output.

Consequently, my new theory of the atomic model is that an atom is actually a galaxy. You must remember though, that the relativity of motion might not show us what a galaxy truly looks like, and as such it might not be a true representation of what an atom looks like either. But it's a start. We see galaxies going in slow motion in the night sky, while we see atoms moving at high speed, but they may be the very same thing, viewed from two different perspectives.

A galaxy, with a massive sun in the centre, or several suns, does not appear to be made of protons and neutrons. So possibly the atoms don't have protons and neutrons, they simply have a nucleus full of electrons packed together, just like a galaxy has a massive sun at its centre, full of the same stuff that it is composed of, crushed solar systems packed together.

However, if there are excellent reasons to continue to believe the nucleus of atoms have protons and neutrons, then the nucleus of any galaxy could be the entire visible galaxy. The massive sun in the centre, along with all the solar systems around it, would be the nucleus. And there would be solar systems orbiting between galaxies, far away from the nucleus.

It would make sense, they always say atoms are empty objects. And away from the nucleus, there are only a few electrons orbiting quite far from the nucleus. If there were a few solar systems orbiting galaxies, far away from the visible galaxies, we might not have noticed. And then, protons and neutrons within a galaxy, would simply be different clusters of solar systems bonded together.

4.5 A solar system is the same as an electron

If galaxies are atoms, and atoms are made of electrons, then solar systems must be the same as electrons. It may seem utterly preposterous, but there is no reason to believe an electron could not look like a solar system. As mentioned just above, no one has even come close to observing an electron. That they may be indivisible insubstantial balls, point particles or made of tiny vibrating strings, are just assumptions.

Hence, my new model of the electron, that they are simpler smaller versions of atoms, composed of neutrinos instead of electrons, with orbiting neutrinos instead of orbiting electrons, could be correct.

This said, the structure of an electron seems less elaborate than the one of an atom, just like the structure of a solar system is much simpler than the one of a galaxy. One is composing the other, solar systems compose galaxies, while electrons compose atoms. They are not exactly similar, and they don't fulfil the same roles. They just have a similar structure of having a nucleus with orbiting particles or objects.

4.6 Suns, planets and moons are respectively the same as super-neutrinos, neutrinos and sub-neutrinos

Suns are super-neutrinos, planets are neutrinos, and moons are sub-neutrinos. Anything smaller, such as comets, asteroids or rocks, are sub-sub-neutrinos. This is essentially my new model of the neutrino. Electrons are made of neutrinos, they have a nucleus with orbiting neutrinos, just like solar systems are made of suns, planets, moons, asteroids and rocks.

This is how much neutrinos can vary in size, and these are all different names neutrinos get at our larger scale. It could be debated that these are all different types of particles, and we could each give them a name instead of just calling them all neutrinos, no matter the size. And indeed, those names are fine. Stars, planets, moons, rocks and grains of sand, then, exist at all scale universes. For convenience, I call them all neutrinos of varying sizes in our very small.

Sub-sub-neutrinos are the smallest particles in our very small scale universe. Just like planets and moons are divisible, and are made of electrons, atoms and molecules, neutrinos must be made of similar particles drastically smaller, belonging to a smaller scale universe. Neutrinos are of variable sizes and are divisible. Smashing neutrinos brings us smaller neutrinos, or sub-sub-neutrinos, just like smashing suns, planets and moons, brings us asteroids and rocks.

Only the particles exchanging smaller particles through orbits between them can be considered charged particles, positive or negative, as only they can either attract or repel other particles. In theory it is true that neutrinos have no charge, positive or negative, hence they are neutral particles. Just like suns, planets and moons must be considered neutral, as they don't exchange smaller objects between them. This said, neutrinos can have magnetic fields made of smaller particles two scale universes down, since suns, planets and moons can have their own magnetic fields made of clouds of electrons.

To what extent these magnetic fields can influence the path of neutrinos must be similar to the influence of magnetic fields on stars and planets, so not much I would say. The paths or trajectories of neutrinos and planets, are guided by the geometry of expansion and the natural orbit effect. No matter the scale, the same laws of physics must apply.

4.7 Rogue solar systems - Proof that galaxies are similar to atoms and form molecules at a higher scale

To prove this theory, we need to identify how galaxies could bond together to form the equivalent of molecules at a higher scale. If atoms exchange electrons in their chemical bonds, then galaxies must be exchanging solar systems. We need to establish how far apart galaxies need to be in the night sky, to be interconnected and form molecules at a higher scale, and it should be proportional to the distance between atoms within molecules. I do not believe they would need to be that close to each other.

And although what we see in the universe is distorted by the relativity of motion, and that we cannot really rely on redshifts to tell us distances, so we can't really be sure if two galaxies are actually near each other, we should still try to observe solar systems orbiting between galaxies. Remember that such rogue solar systems could be quite far from the entire visible galaxies, which are just the nuclei of the galaxies.

There should also be several solar systems orbiting the galaxies themselves, but on much higher orbits and far from the visible galaxies, possibly extending to the midpoint between galaxies. So, solar systems orbiting galaxies, orbiting between galaxies, and orbiting between clusters of galaxies. We should start a hunt for any rogue solar system out there, which is outside any galaxy.

It is not going to be easy. Almost all of the stars we see in the night sky with the naked eye, if not all of them in fact, are just the ones within our own galaxy the Milky Way. It is at least proven that there are rogue stars outside galaxies, possibly in some cases without planets orbiting them. It would not matter if there were planets or not around these stars, they could still bond galaxies together.

What I am talking about is more a matter of course. There would be rogue stars and rogue solar systems everywhere, as part of the formation of galaxies. It would not be just an unusual occurrence, due to some galaxies crashing into one another, or galaxies imploding and exploding, spewing a few rogue stars out there, as is currently thought, in order to explain rogue stars found outside of galaxies.

If we assume that our galaxy, the Milky Way, is forming a molecule at a higher scale with our nearest neighbours, the local group of 30 or so galaxies, then we get some idea of how far apart galaxies can be, while still being bonded together by orbiting solar systems. These solar systems that are not within galaxies have a name, according to Wikipedia they are called intergalactic stars, or rogue stars, and they were only discovered in 1997.

Already we have identified 675 intergalactic stars, and probably solar systems, at the edge of our galaxy, and between the Milky Way and the Andromeda galaxy. There are probably many more that we have not yet seen. Studying the trajectory of their orbits is what we need to do next. My theory of the model of the atom as being a galaxy could be verified, if we could prove or observe that these rogue star systems are in fact orbiting constantly between both galaxies, and other galaxies, creating a chemical bond between these galaxies, which could then be part of a molecule at a higher scale.

We know galaxies exchange magnetic fields, however, that is made of the stuff at the lower scale, clouds of electrons. It could not bound larger scale objects together. If galaxies were not linked through solar systems orbiting between them, and they were linked instead through magnetic fields made of electrons, it would mean atoms are bonded together through magnetic fields of particles at an even smaller scale, instead of electrons. We know this is not the case.

Maybe atoms exchange magnetic fields of particles similar to clouds of electrons, but at a much smaller scale than our very small, but it could not help to create any kind of positive or negative charge, or attracting or repelling force, it would be too weak. Just like the Earth's orbit has nothing to do with its magnetic fields, since some planets don't have any magnetic fields. And of all the moons in our solar system, apparently only Ganymede has a magnetic field. The exchange of electrons between atoms is how they bond together and form molecules, and so galaxies must be exchanging solar systems between each other.

We must assume of course that those galaxies are indeed not just atoms lost in space, and are actually part of molecular structures at a higher scale. It is safe to assume that any grouping of galaxies close to each other must form cosmic molecules. But even different groupings of galaxies very far away, could also be part of that same molecular structure at a higher scale, and they also would somehow be exchanging solar systems between them from that far apart.

For example, if all our nearest galaxies form one molecule, then another cluster of galaxies further away would form another molecule, and often all molecules are bonded together through what must be electrons. Hence, solar systems must also be orbiting between clusters of galaxies in order to bond them together.

I mentioned earlier that Mark McCutcheon and Standard Physics could not see much in common between an electron and a solar system. Some of these reasons are still valid when we compare atoms to galaxies. How could a galaxy achieve anything atoms appear to be achieving at a smaller scale? For example, can galaxies be squeezed, can they push back other galaxies? And what else these atoms and galaxies do, or don't do, have we identified?

The question is, are we sure atoms are doing exactly what we think they are doing? We are still, to this day, unable to have a good look at them. And the day that we can finally take good photos of atoms, they are moving so fast, it will be hard to tell what is going on exactly.

Are we certain we have seen the extent of what galaxies can do?

They don't move very fast from our point of view. What we see is more like a photo shoot of different galaxies, and from that we are to piece together their movements and how they are interconnected. Our actual methods of assessing their distance are not even reliable, as you can read in the article included in this book called: *Dark-Matter, Dark-Energy and the Big-Bang All Finally Resolved.* And also the free excerpt from *The Final Theory,* available on my website *www.themarginal.com* called: *Cosmology in Crisis.* I would say that we are far from being able to reach any kind of conclusion on the subject.

It is still possible that, because the laws of physics in Atomic Expansion Theory can be applied to both scales, smaller particles in the lower scale would shape up into larger similar structures at the higher scale, but there would end the similarity. No cosmic molecules then, and both the atom and the galaxy would simply be similar, because the same laws of physics apply to both, no matter the scale. They follow in their course the geometry of expansion and the natural orbit effect. However, if the laws of physics are the same at both scales, why would something extremely similar at a higher scale acts any differently than it does at the lower scale? As above, so below; as below, so above.

4.8 Rogue planets - Proof that solar systems with orbiting planets are similar to electrons with orbiting neutrinos

A binary star system is a solar system where planets are orbiting between two stars, and so on for multiple star systems where planets orbit several stars. Although we have never directly seen yet any planet orbiting other suns, or barely, as we can only see the multiple suns in these solar systems, it is assumed to be correct. This said, I think we have by now seen black dots over other suns, which must be planets.

This could be similar to several electrons bonded together to form a larger particle like a proton or a neutron. Yet, it would not seem to be enough, as such multiple star systems would only represent but a few electrons bonded together, when we know protons and neutrons are made of many more electrons. It is more likely that a multiple star system is still just one electron, not several electrons bonded together.

Mark McCutcheon explains these electrons staying together, by the simple force of expansion, as they push each other in their expansion. But we could surmise instead that they are attached through the exchange of orbiting neutrinos between them. At the larger scale then, distant solar systems, no matter how many stars they contain, must also be connected through rogue planets orbiting between these solar systems.

A bit like a molecule is formed of atoms, but instead these electrons and solar systems would simply be forming larger particles at their own scale. In the case of electrons, they would be forming any larger particle like proton, neutron, clusters in light and heat, or clouds in magnetic fields. This is what solar systems bonded together must be forming at a higher scale.

Therefore, although a galaxy appears to be filled randomly with a whole lot of solar systems, several could be connected together and be the equivalent to protons and neutrons within the atom. In that particular scenario, the entire visible galaxy would be the nucleus, with the orbiting rogue solar systems being nearly unseen, and at an astronomical distance from the entire galaxy.

To prove my theory, we need to identify planets orbiting in higher orbits, very far from our solar system, further away than our Sun and our familiar eight planets, which form the nucleus of our solar system. Such planets would be extremely difficult to detect, since they don't emit light, and never come close to us. The rogue planets orbiting between solar systems would also take an extremely long time to travel from one solar system to another.

We have already identified such rogue planets in the nearby systems. We might not have calculated their orbits yet, but I think we will find that they must either be in very large orbits around the sun, or orbiting between solar systems, connecting nearby solar systems together.

It was recently calculated that there must be another planet as yet unidentified in our solar system, and possibly it could be that this planet is actually orbiting between our Sun and one of the nearby solar systems, like Alpha Centauri, Bernard's Star, Wolf 359, Lalande 21185, and finally Sirius.

Although, it is more likely that if we can detect some orbit anomaly within the planets of our solar system, then this missing planet is not rogue, and is not orbiting between solar systems. It just has a very large orbit around our Sun, and probably not an exceedingly large one at that. Otherwise, it would have little or no effect on the orbits of our eight known planets. So that missing planet might be part of the nucleus.

The famous Planet X, also known as Nibiru, has been described as originally orbiting Sirius C, but that now it is part of our solar system, following a long orbit of perhaps 3600 years around our Sun. Maybe it is actually still orbiting between Sirius C and our Sun, creating a bond between both solar systems, just like neutrinos are creating bonds between electrons.

Would an orbit of 3600 years be enough time, for a planet to go from one solar system to another, and back? It is possible that even though that planet came from Sirius, and was orbiting for a while between both our systems, due to the natural orbit effect of the geometry of expansion, after hitting a few things, it no longer has the escape velocity to reach Sirius. Who knows? Who knows even if this planet really exists? Then perhaps other planets, that we are still unaware of, are doing that journey between our star and Sirius, and also other planets between our Sun and the suns of other solar systems nearby.

It is not a big stretch to imagine how planets could come to orbit two solar systems that far apart. Just like it can easily be imagined that solar systems could be orbiting between two galaxies, also extremely far apart. They would follow tremendously large and long orbits around two more massive expanding objects, and follow the natural orbit effect of the geometry of expansion.

If these planets are orbiting near the midpoint between two solar systems, and if there is a depletion of planets orbiting one of them, then in search for balance due to the geometry of expansion, these planets could easily start orbiting between both solar systems. The same for solar systems. If they are orbiting in higher orbits near the midpoint between two galaxies, and if there is a depletion of solar systems orbiting one of the galaxies, then solar systems could easily start orbiting between both galaxies. And the same for solar systems orbiting around and between clusters of galaxies.

This would create attracting and repelling forces, keeping the solar systems apart and the galaxies apart. It does not mean that it is fool proof, and that these smaller objects orbiting these larger ones, would totally prevent galaxies from colliding with each other, and the same for solar systems. However, it is safe to assume that for most of the time it works, and more pressure or objects added to any system, would simply affect all these objects by changing their speed and orbits, pushing and pulling galaxies together, pushing and pulling solar systems together, so they don't crash into each other. In that way, galaxies and solar systems can be squeezed, and they can push back, vigorously, all through the natural orbit effect, and through the geometry of expansion.

4.9 Figure 8 shaped orbits for smaller particles or objects, orbiting at the midpoint between larger ones

After visualising orbits of these smaller particles or objects bonding larger ones together, it occurred to me that there could be another kind of orbit for them, near the midpoint between the larger ones. It is the one of an 8 shaped orbit. And it would also apply to planets that would be orbiting two stars or more, in multiple star systems.

I read an article in the *Sky at Night* Magazine, published by the BBC on 23 October 2020, titled: *What happens when exoplanets orbit two stars at once?* In binary stars systems, it is said that "a planet can either orbit in a 'satellite-type' or S-type internal orbit, whereby the planet circles one of the two stars". Or, "the planet could have a wide, circumbinary orbit around both stars in the middle - what's known as a 'planet-type' or P-type external orbit". It is also thought that there could be a third kind of orbit: "It's even theoretically possible for a planet to be held within the L4 or L5 Lagrangian equilibrium point, co-orbiting with the smaller star around the larger star, which is comparable to how the Trojan asteroids sit within Jupiter's Lagrange points in our Solar System."

This is also stated on the Wikipedia page called *Habitability of binary star systems,* from where I reproduced and amended an image by Philip D. Hall:

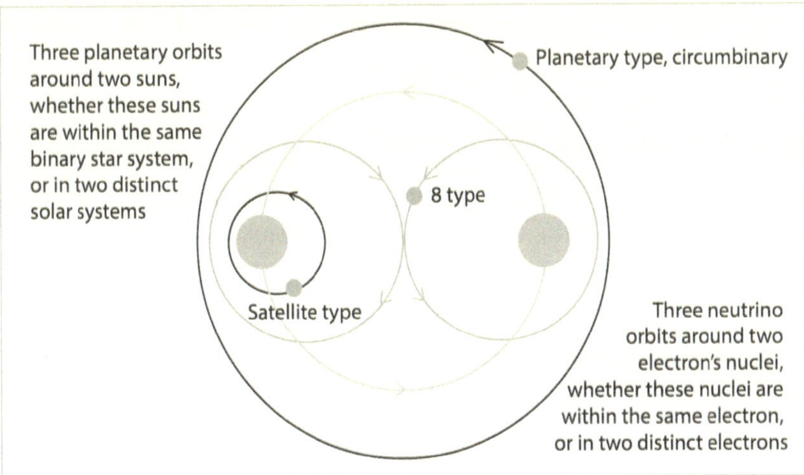

Figure 17 Three different planetary orbits around two suns, whether these suns are within the same binary star system, or in two distinct solar systems. P-type orbit (Planetary-type, circumbinary), S-type orbit (Satellite-type), and 8-type orbit

Now, imagine that another planet is orbiting the right star, but nearer the centre of both. It is unlikely that an orbit around one star, could turn into a stadium or sausage shaped orbit, if suddenly the planet were to start orbiting both stars. Since the planet would already be orbiting in a circular or elliptical orbit around the right star, it probably could only switch to orbiting the other star around the midpoint. And from there on, the planet might only orbit the left star, or it could continue to orbit both stars in an 8 shaped orbit.

Moreover, while the planet would be orbiting counter clockwise around the right star, once it switches to orbiting the left star, it would be orbiting clockwise. And, in the geometry of expansion, as well as the number of planets orbiting each star, this change of direction of the orbit, might play a role in attracting or repelling other stars.

It is the same principle for planets orbiting between solar systems, solar systems orbiting between galaxies, and solar systems orbiting between clusters of galaxies, all near their midpoint. Transpose that to the very small, and we find that it is the same for neutrinos orbiting between electrons, electrons orbiting between atoms, and electrons orbiting between molecules, also near their midpoint.

It seems logical to me, that there could be smaller particles or objects orbiting near the midpoint between larger ones, and that their orbit between the two particles or objects, would be 8 shaped. Half the time their orbits would be clockwise, and the other half, they would be counter clockwise. I believe the idea of 8-type orbits within binary systems is not new, but somehow it was not possible within Einstein's theory, which we are now freed from.

And this could potentially play a role in the attracting and repelling forces in Coulomb's Law in electrostatics, which I will discuss in the next chapter. It could also play a role in magnetism, in the formation of larger particles made of bonded electrons, and in the bonding of atoms forming molecules. In any case, I would still expect other smaller particles or objects, to orbit the overall larger particles, as in the planetary-type, circumbinary orbit, mentioned above.

I won't be talking about these 8-type orbits again, since I cannot be certain. But anyone attempting to track the orbits of rogue planets and rogue solar systems, should keep this in mind. Computer simulations involving expanding matter, should be able to tell us if the 8-type orbits are likely or not.

Please note that these 8 shaped orbits are not to be confused with analemma, which is also an 8 shaped orbit, for example of the Sun, as observed in the same location on Earth over a year period. Analemma is clearly not a real figure 8 shaped orbit, it only appears to be one, from that one location on Earth, while the Earth is orbiting around the Sun.

4.10 What could the universe form at a larger scale?

Assuming that galaxies are atoms, that they connect with other galaxies to form molecular structures at a higher scale, then what could this molecular structure be? What is the universe forming at a higher scale? You can search online for mapping of the universe, checking what clusters of galaxies might form.

The latest mappings of the universe, from *The Millennium Simulation Project* at the Max Planck Institute for Astrophysics in Germany, are computer simulations. They are based on redshifts, to measure how far away the galaxies and galaxy clusters are, and on mapping dark matter. This makes it all completely inaccurate, since there is no dark matter to be found in the universe, no dark energy either. Moreover, redshifts cannot be relied upon to measure neither the distance of objects in the universe, nor their speed, nor their direction of motion. See the article and excerpt mentioned two sections ago.

Still, people have seen in these maps, the neurones or synapses of a brain. Or the branches of a tree, or potatoes. Take your pick. No matter how biological the universe might look like, the truth however is that it could be anything. At the very least, we might be seeing molecular structures, galaxies interconnected together, and located in certain locations, in order to form that structure at a higher scale.

Let's try to assess the real scale of the observable universe, if we were to bring it to our own scale, assuming galaxies are in fact atoms. I checked online several sources, and although it varies greatly from one source to another, it appears that there could be between 20 and 100 trillion atoms within one human cell. At the larger scale, and although this number is growing as we get more powerful telescopes, we are seeing about 200 billion galaxies in the observable universe.

No matter if these numbers are off the mark by a large margin, we can see that the entire observable universe, at our scale, would be much smaller than a human cell. The human body would contain as many cells as there are atoms in one cell: 100 trillion cells. We really see nothing of the higher scale universe. Possibly the entire universe is, as my psychic friend in Los Angeles saw in a vision, a cell. And then, the universe would continue further away in other cells, so there is no end to it.

God only knows what a human cell would look like if we were to look at it from the very small scale universe, from the point of view of atoms. But it could look precisely like the images of the universe above. Especially because, as I stated before, our methods of establishing how distant galaxies truly are, are flawed, relying on redshifts, as if light was like the Doppler shift effect in sound, when it is nothing of the sort.

Maybe we made countless mistakes in assessing exactly where these galaxies are in the universe, and they may not be interconnected as we see them in these images. Also, the relativity of motion distorts what we really see out there. Possibly then, the molecular structures we are seeing in the night sky could be DNA and RNA structures, or other smaller stuff as contained within a human cell, or it could be anything else that is not biological or organic.

Interestingly, a cell is another such object that has a nucleus with floating things around it, for a better word than orbiting. Just like the original Plum Pudding model of the atom, where electrons were thought of as raisins floating in some pudding, but with no nucleus at all. And for a while, I thought the model of the atom could be very similar to a cell. They would divide and multiply the same way, while still expanding and growing. I am still working on this theory. For now though, I believe that the best theory is the one I am stating in this present book.

To conclude, there is no reason to believe the universe could represent a cell at our scale. I only used this as an analogy to give us an idea that the observable universe represents only a small fraction of the size of a cell.

We have no idea really how far the universe goes. Our numbers at the moment rest entirely on how far we have been able to see. And physicists believing in the Big Bang, which is not required in Atomic Expansion Theory, believe the universe has an end, when perhaps it does not. Or if there is an end to the universe, no matter how great the future telescopes we could build, we might still only ever see an extremely small fraction of it.

I must point out that there are life forms made of one cell only, they are called single-celled organisms, such as an amoeba and other bacteria. The universe could indeed be self-aware and alive, but not the kind of intelligence we would have expected. Unless of course there are trillion other similar universes out there, each forming a cell, and together they form something more intelligent at a higher scale. As I said though, maybe it does not form something which is alive or organic. We don't know, despite the fact that Seth did say in *The Early Sessions,* that we were like cells, a part and forming a larger entity at a higher scale.

FIVE

The speed of light is variable and subject to gravity - Gravity versus attracting and repelling forces - Newton's Law of Universal Gravitation versus Coulomb's Law

5.1 The electromagnetic spectrum in Atomic Expansion Theory

In *The Final Theory,* Mark McCutcheon presents his new electromagnetic spectrum according to Atomic Expansion Theory, which is now separated in two clearly defined sections. The first half goes from thick electron compression bands starting with AM radio, to thin bands of compressed electrons, ending with microwaves.

The second half however, is made of electron clusters, from large clusters in infrared light, to small clusters in gamma rays. Visible light immediately follows infrared light and is made of large electron clusters.

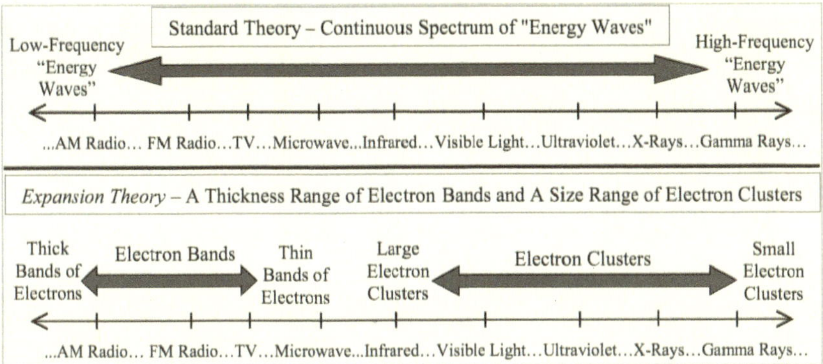

Figure 18 (5-6) Electromagnetic spectrum today vs. Atomic Expansion Theory

Higher light intensity, means more clusters within the light beam. And a higher frequency, means the clusters are simply smaller, so verging more towards gamma rays, towards the end of the electromagnetic spectrum. Matter vibrating at a higher frequency would therefore be outside our field of vision, outside visible light.

McCutcheon states that the different sizes of these clusters are not referring to the real physical size of the clusters, but instead to the number of electrons within these clusters. He says that electron clouds and electron clusters are not atomic objects, and they have no true size definition in the atomic realm.

If you ask any scientist today, what is the size of a photon of light, you will get an answer that its size is more related to its frequency or its wavelength, since photons are considered to be acting more like waves than particles. And if you ask them about the size of electrons, the answer will be that they are considered to be point particles, with no size nor volume, and considered to have no mass. There is a debate today in science, concerning the possibility that photons might have a mass. With these kinds of definitions, one could actually wonder if subatomic particles and light truly exist.

It stands to reason, that the more electrons there are in any cluster, the larger is the cluster of electrons. It also stands to reason, that the electron clusters not only have a size, but also a volume and a mass. In electrostatics, subatomic particles have charges, but it simply means that the more charged a particle is, the more electrons there are within it or on its surface. So again, we come back to size. Something is expanding, into a sizeable thing at this kind of rate, reducing distances between such subatomic particles.

It is not only the speed of light that we need to consider, it is also the speed of any compressed bands of electrons, as well as any size of clusters of electrons. All these types of bands of electrons and electromagnetic radiations, in the figure above, are said to travel at precisely the speed of light.

If you remember *Today's energy terms according to Expansion Theory* in chapter 3, section *3.9 Energy*, we also need to be able to measure the speed of clouds of electrons which form magnetic fields, and the electrons within the atomic realm, and the exchange of electrons in chemical bonds. The whole branch of electrostatics, governed by Coulomb's Law, needs to be considered.

If such particles were not moving so fast, and were not so small and invisible to our instruments, that we only detected their behaviour through energy levels and energy exchange, or frequencies and wavelengths, it would be an entirely different story. If only we could see more clearly what is truly going on at that scale, it would help. I have already discussed, how the very small is identical to the very large universe, and how it can help us figure out the truth at the smaller scale, once we compare both scales. Let's keep this in mind for the rest of this chapter.

5.2 Subatomic and atomic expansion rates are the same - The speed of light is variable

According to Mark McCutcheon, the expansion rate of the electron is much higher than the one of the atom. And unlike for the Universal Atomic Expansion Rate X_A, which is known, there is no way to calculate what the Universal Subatomic Expansion Rate X_S might be. The higher expansion rate would explain how the electrons can escape the atoms, and why the speed of light is so high, which depends on that higher expansion rate of the subatomic particles.

I must admit, it is very neat to have the speed of light depend on the expansion rate of the clusters of electrons within light or heat. As the clusters move through space, they are actually simply expanding at an incredible constant rate, pushing against each other, justifying their motion and the constancy of the speed of light. And since they would simply expand in all directions from their point of origin, I used to think that it would also explain two things from Quantum Mechanics.

In the double slit experiment, by the time the expanding electron reaches the two slits, it could have expanded enough to go through both slits at the same time, causing its own interference pattern on the back screen. And quantum entanglement would be easier to explain, if these electrons were simply expanding against each other, from their point of origin, to wherever they go, remaining interconnected the whole time during their expansion. Then of course, changing the polarity of one particle, would instantly change the polarity of the other particle, which might be miles away, but in fact is still touching the first one. But I had to dismiss these explanations for the reasons I will soon describe. The right explanations for the double slit experiment and quantum entanglement, according to Atomic Expansion Theory, have already been covered in two sections near the end of chapter 2.

For this to work, we would have to accept that electrons can expand so much, that they would quickly outgrow entire solar systems and galaxies, and even the entire universe by the time they reach any assumed edge of it. Indeed, if a star goes supernova, in time it is seen at astronomical distances. Can the electron clusters within light expand to that size, even if according to Mark McCutcheon and Standard Physics, electrons don't have any real substance or tangible reality to speak of?

There is another obstacle when it comes to my own theory, where electrons are actually solar systems. What one does, the other must also do. Solar systems could be launched out of the galaxies at high speed, if the entire galaxies were to explode, just like electrons could be launched out of the atoms, if they were excited enough, and the atoms broken down.

But solar systems could not expand forever at high speed, their inner expansion rate would have to remain constant, as per the first type of expansion of matter. And from the second type of growth and shrinkage of matter, in the crossover effect, there would have to be a limit to the enlargement of the planets' orbital rings around their sun or suns, and to the size of the orbits of the planets orbiting between solar systems. Otherwise, the entire structure would lose coherence and fizzle out in all directions.

It is more likely that electrons are pushed out of the atoms at high speed when the atoms collapse, which is like a galaxy exploding and sending all its solar systems in all directions, while their speed and motion are not dependent on their inner expansion rate. In fact, I feel that the Subatomic Expansion Rate of the electrons, is the same as the Atomic Expansion Rate of the atoms. Which seems logical, and would explain why a static electric field, can remain absolutely static, while everything else expands at the same relatively slow constant rate.

We would need to consider a few changes from Mark McCutcheon's theory, in order to accept the idea that electrons expand at the same rate as atoms. The first change, the electrons are not bouncing on the nucleus of the atom, they are orbiting. The second one, the inside of the atom is not a foreign dimension that exists outside of our reality. The third one, is that the electron could have a nucleus with orbiting neutrinos.

It is this exchange of orbiting neutrinos between the electrons, that creates this attracting and repelling force that has been described as a charge in Coulomb. This exchange of neutrinos also creates the bonds between the electrons, in order to form stable particles such as protons, neutrons, clusters and clouds.

In Mark McCutcheon's description of the crossover effect, he said that the expansion rate of the electron must be faster than the one of the atom, radically faster, otherwise, just like rocks sitting on Earth, the electrons could never leave the surface of the atom. And this is despite their bouncing behaviour. Indeed, why would they suddenly start bouncing faster and higher, even if there were to be suddenly more of them around?

But the electrons could leave the atom easily, if they were instead orbiting far from the nucleus, and if they had an extra boost from the expansion of the orbits of the neutrinos around and between them, pushing them out as the neutrinos accelerate in their orbits. I believe that this is how electrons attract and repel other particles, and how they move around, through this exchange of orbiting neutrinos.

Also, for electrons to be pushed out of the atoms, when it comes to a large number of them, which must come from the nucleus, the atoms might need to collapse or explode, sending all these electrons out at a very high speed. There would be no problem then, to explain how they can exit the atoms.

Inside the atom, the orbits of the neutrinos within and between the electrons, would be extremely small, bringing the neutrinos nearer the nucleus of the electrons, causing a shrinkage of the electrons. But as the electrons exit the atom in the crossover effect, these orbits could be enlarging considerably, as the neutrinos are excited and start to orbit faster, both inside the electron and between the electrons. And this is the push the electrons need to leave the atom.

In my opinion, this is also the cause of the second type of the growth and shrinkage of the electrons in the crossover effect. When inside the atom, the electrons are kept under restraint and apart, by the other electrons orbiting around, and their exchange of orbiting neutrinos. But once outside the atom, the neutrinos are suddenly free to orbit faster, causing the orbits to enlarge, and the electrons to grow. Even although the orbits of these neutrinos can keep a steady speed and size, as in a static electric field.

In static electricity, when two clouds of growing electrons of the same density meet, between two positively charged rods of glass, and they could repel forever if left untouched, McCutcheon states that the inner expansion of the electrons is still ongoing behind the scenes. But in order for the whole setup to remain completely immobile, the inner expansion rate of the electrons X_S, must be identical to the one of the atoms X_A. Otherwise, it would never remain static, the entire cloud of electrons would quickly expand faster than everything else around, especially the two rods of glass made of atoms and molecules.

In order to explain how the electrons could be expanding so fast at all times, whether they are inside or outside the atom, while not causing the atom to also expand greatly, Mark McCutcheon stated that the inside of the atom is a foreign dimension, and is not part of our universe or reality. He says that our reality exists because of the space between the atoms. And so, electrons can still be expanding greatly within the atom, without causing the atom to proportionally expand at such a high rate as a result. This however does not take into account that in a static electric field, the electrons are very much already outside the atoms, and outside that foreign dimension. And yet, their inner expansion rate cannot be that high, otherwise the field would not remain static.

If we move away from the idea that the inside of an atom is a foreign dimension to our reality, then the only way the electrons could be prevented from quickly outgrowing the atom, would be if the expansion rate of the electron was identical to the one of the atom. Since, how could smaller particles, composing a larger one, could expand faster than the larger one they are composing? Unless atoms are like Doctor Who's Tardis, they are larger in the inside than they appear from the outside. Was the idea of a foreign dimension within the atom invented, just because an explanation was required?

It is more sensible if the neutrinos orbit the electrons, and orbit between them, in order to bond them together. With the orbits of the neutrinos getting larger or smaller, depending on their speed and encounters with other particles. This causes the growth and shrinkage of the electrons in the crossover effect, and the attracting and repelling forces of Coulomb's Law. And then, the constant inner expansion rate of the subatomic particles behind the scenes, could be the same as the atomic one.

I cannot think of any other argument, as to why the expansion rate of the electrons should be faster than the one of the atoms, except one. The speed at which freely expanding electron clusters are going at in space, the speed of light. Why would it be so high? This would suggest a very high expansion rate, for those electrons and electron clusters, compared with the atoms' expansion rate. But does it?

Just like when electrons are ejected from the atoms through the crossover effect, the only way they can be completely freed from the subatomic realm, is when they are very excited, with all their neutrinos orbiting faster, with orbits extremely enlarged. This pushes them out into space at quite a speed once freed, while the inner expansion rate of the particles composing the electrons does not change, and could still be the same as the one of the atoms.

The only way we could get enough electrons to form clusters, or clouds, would be if we use the electrons within the nucleus of the atoms. So quite possibly, the atoms must collapse or explode, before any cluster or cloud of electrons can form. Then, certainly the electrons would exit the atoms at a very high speed.

After they are freed at high speed, the speed of these electron clusters is more like gravity, a changing distance between expanding subatomic particles, and whatever else we are trying to measure their distance from. If the expansion rate of both the electron and the atom are the same, then the Atomic Expansion Equation to calculate gravity, applies to both atomic and subatomic objects. Which is why I renamed it the Atomic *and Subatomic* Expansion Equation.

People will think I am a total lunatic for saying that the speed of light is variable and accelerating, despite all the evidence and proofs out there to the contrary. Well, I say it is time to check again, and find better ways to measure the speed of light over very long distances.

I swear I can even see it with the naked eye. When a spotlight is lit up in the valley where I live in Wales, it seems to me that light is slower at the start, and suddenly accelerates towards the other end of the valley. But I could be imagining it. After all, it is most likely that if we failed to realise that the speed of light was accelerating, it is because our measurements here on Earth might be insufficient to assess it. Or, we never considered the possibility, so we never tried to measure it before. I'm not sure which it is.

Because, what is the difference between clusters of electrons and asteroids hurtling into space? Both are expanding objects, and the changing distance between them, and for example Earth, should be a calculation of gravity, a changing distance between expanding objects. And it should be an acceleration, because of the two distance decreases to be considered when calculating gravity.

Clusters of electrons within light passing by the Sun, are acting very much the same as a passing comet. Both their paths are affected by the expansion of the Sun and the geometry of expansion, and they enter into a partial orbit around the Sun, before continuing on their new trajectories. Light is moving faster than the comet, because originally the clusters were freed at a much higher speed.

Whatever way we look at it, the speed of light must be accelerating, just like for a free floating object being caught up by the expanding Earth. In the end, I believe clusters of electrons are objects too, they have a true definition, a true size, a mass and a materiality. They are no longer photons or packets of energy without form, mass, definition or reality.

Subatomic particles freely expanding into space cannot simply bypass gravity, just as they cannot bypass the geometry of expansion. Light might never enter into a full orbit with anything, as it is going too fast, but certainly its path deviates when passing larger objects, as seen in the Gravitational Lensing effect. Which means, the clusters of electrons in light enter into a partial orbit, before continuing on their way, and are accelerated by the Slingshot

Effect, just like probes we send into partial orbits around large planets, in order to accelerate them.

If the speed of the clusters of electrons was entirely due to the high expansion rate of the subatomic particles, then the acceleration of the distance decrease between them and their destination, or gravity, would be extreme. Because the distance decrease depends on the expansion rate of the subatomic particles, an expansion rate that is very high and unknown in *The Final Theory.*

Light is fast, but not that fast. It cannot be accelerating at an incredible rate with distance, as it would according to the Atomic and Subatomic Expansion Equation in gravity, if its expansion rate went ballistic. Consequently, the electrons' expansion rate cannot be radically higher than the one of the atoms.

Since most of the speed of the electron clusters is due to the process used in order to free them from the subatomic realm, for example through heating the element of a lightbulb, and that they are launched at high speed as a consequence, then the acceleration of the distance decrease in gravity is manageable. Because then, the expansion rate of the subatomic particles does not go ballistic, it is the same as the one of the atoms. Like everything else, electrons double in size every 19 minutes, instead of perhaps every few seconds. Hence, the speed of light does not accelerate at an alarming rate, which would make the whole idea of the accelerating speed of light untenable.

Note that in the scenarios described above, electrons are all more or less the same size and equalised. It is only the orbits of the neutrinos within them, and between them, bonding them together, that change size with speed, causing the growth and shrinkage of the electrons and whatever they compose, such as clouds and clusters. The second type of growth and shrinkage of subatomic particles, is completely independent from the inner expansion rate of all matter, the first type, which is constant and ongoing no matter what.

The electrons equalise in size when in contact with other

electrons, in their search for balance, which is entirely due to the natural orbit effect of the geometry of expansion. To visualise this at the larger scale, if several solar systems of different sizes where to get close to each other - assuming they had a different number of planets, and that they would attract instead of repel - the exchange of planets between them would eventually bring standard size orbits between all the planets of all the solar systems. All the solar systems would end up being of similar sizes. Just like when two clouds of electrons of a different density, meet between a rod of glass and a rod of plastic, the electrons are balancing to reach the same size, reducing the distance between the rods, creating the appearance of an attracting force.

That way, the electrons can have a real true definition and size whether they are within or outside the atom. It is just that within the atom, the orbits of the neutrinos around the electrons' nucleus, are moving slowly and are shrunk, while once outside the atom, the orbits are faster and enlarged.

In the crossover effect, within the electron, the electron's nucleus and the neutrinos themselves, are not suddenly expanding faster, just like we would not expect our Sun and planets to suddenly expand faster. They are still expanding at their inner constant expansion rate, which in my opinion must be the same for all of these particles and objects, whether they are molecules, atoms, electrons, neutrinos, galaxies, solar systems, planets or suns.

The electrons and neutrinos are orbiting instead of bouncing, because everything I have just stated above works exactly the same in the very large. That is, if you consider that neutrinos are planets, electrons are solar systems, and atoms are galaxies. Nothing is bouncing at the larger scale.

At this time, it might be hard to visualise everything I am stating here, but through reading, absorbing and visualising what I am saying, eventually we can see that it all works perfectly. There is nothing now in the very small which we cannot also observe in the very large. In fact, comparing the two scales is how we can obtain many answers, and discover new phenomena we would never have thought of looking for previously.

Having the electrons suddenly having a nucleus, with orbiting and exchanging neutrinos, explains the growth and shrinkage of the electrons within the crossover effect. These are compelling arguments to justify how the Subatomic Expansion Rate could be the same as the Atomic Expansion Rate.

This is no small thing, it is a game changer for our equations and calculations in physics. Because then, we know the inner expansion rate of the electron, which was one big unknown variable in Mark McCutcheon's *The Final Theory*. And if we know the expansion rate of the subatomic particles, then we can use the Atomic and Subatomic Expansion Equation, to calculate the speed of light and electrons accurately.

We can also now calculate the effects of gravity at that small scale, as long as we have a reasonable measurement of the size of subatomic particles, and of the electrons. We could even find out the size of subatomic particles and electrons, if all the other variables of the Atomic and Subatomic Expansion Equation were known.

To measure the speed of light, we need to first assess the initial speed at which the clusters of electrons are expelled freely into space, which will always account for most of the speed of light. And from there, we use the Atomic and Subatomic Expansion Equation, to measure the additional speed gained through gravity, the changing distance between expanding objects. And then, we need to take into account slingshot effects that accelerates light as it passes near large bodies, and anything light might encounter in space that might slow it down, such as gases and magnetic fields.

To conclude, the speed of light must be variable, and it must constantly be decelerated and accelerated as it travels. Charting the speed and path of clusters of electrons in space, is the same as charting moving asteroids. And electrons freed from the atoms and the subatomic realm, meaning outside the atoms but still interconnected to particles not freely expanding into space, is the same as a gun shooting bullets. Any thought experiment involving the speed of light, should keep this in mind, to help visualise exactly how these electron clusters will behave.

And now, everyone can see light coming towards them, no matter if they are going at a normal speed, or at multiple times our previous assessment of the speed of light. In all cases, it is just how long it takes for expanding objects to reduce the distance between the point of view and the origin of the light source, while considering the initial speed at which the clusters in light were launched into space. There is no such thing anymore as a maximum speed in the universe, or any set speed for light. Let's not forget, motion is relative to the observer's point of view, including the motion of light. I will discuss in chapter 7, how it is the speed of light that is relative instead of time, when I discuss Special Relativity and relative motion.

5.3 The varying speed of light and the size of the clusters of the electrons (light frequency) in gravity

In the previous section, I discussed why the speed of light accelerates due to gravity and the geometry of expansion. In this section, I am addressing all the ways by which the speed of light can be affected and vary in its travel, to offer the best ways possible to measure it, depending on the circumstances. It can get complicated, as there are many factors to take into consideration, and some unknowns as well.

The electrons cannot grow that much in the crossover effect, it is unlikely they would grow larger than atoms. Since it would mean that a solar system's orbits could expand larger than a galaxy. It is more likely that they only expand to the size of a few electrons, while the neutrinos in their orbits accelerate. It could be enough for the electrons to escape the atoms. We are very likely talking about electrons orbiting very far from the nucleus of the atoms, or solar systems orbiting very far from the visible galaxies.

Mark McCutcheon mentions that when it comes to the electrons' expansion rate, we should consider the number of electrons within the electron clusters, instead of their size, since true size does not truly apply to subatomic particles, as he feels they don't really have a material existence like atoms.

Whether we think in terms of the number of electrons within electron clusters, or their size, does not really matter. It stands to common sense, that the more electrons there are, the larger the clusters must be, the more mass they have, and the more charge they have. It is all an equivalent way of comparing quantities in our equations, and it is why McCutcheon, Newton and Coulomb, all developed equations which sometimes can be interchangeable, and can bring similar results. I will review these equations later in this chapter.

I am referring here to the components of the electrons, which are their nucleus and their orbiting neutrinos. As mentioned before, their expansion rate is constant, and the same rate as the one of the atoms. Electrons' components don't suddenly expand faster in the crossover effect, although they cause the electrons to grow and shrink. It is only the orbits of the neutrinos that expand or shrink, rendering electrons able to attract or repel other particles, through an interchange of orbiting neutrinos between them, linking them together, and creating what is referred to as a charge.

The electrons never expand that much, even while travelling and expanding freely into space. The growth of electrons due to the crossover effect, the enlargement of the neutrinos' orbits within and between electrons, cannot either make electrons grow crazily, since there must be some sort of limit before they lose cohesion and the neutrinos are lost in space.

In Atomic Expansion Theory, the lower the frequency of light, the larger are the clusters of electrons, such as in infrared light. And the higher the frequency, such as in gamma rays, the smaller are the clusters. When we talk about clusters of electrons, large or small, Mark also says that in a given length, there would be the same number of electrons no matter the size of the clusters, as can be seen in figure 19. Consequently, their speed, which depends on their expansion rate, would be the same, which is the speed of light.

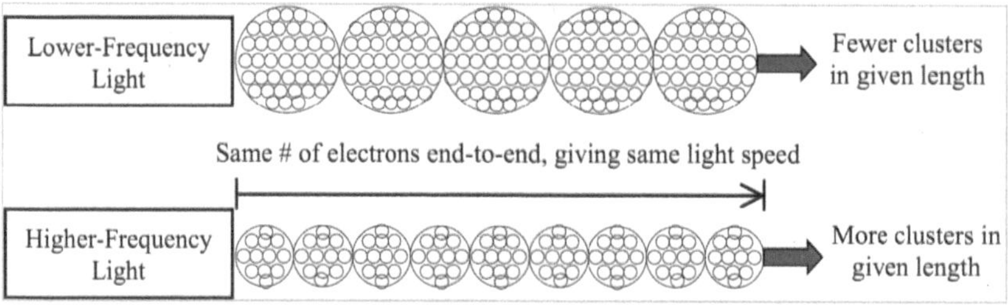

Figure 19 (5-4) Two light beams: same speed, different number of clusters

McCutcheon says the size of the clusters has no definition, it refers instead to the number of electrons within the clusters. Clusters of electrons can expand a great deal as they travel, but it does not change the frequency of light, or anything else, since it is the number of electrons that counts. For McCutcheon, electrons only have definition within the subatomic realm. However, I disagree.

The size of an object we use, in the Atomic and Subatomic Expansion Equation, must be the size of the object as it is, not the number of them. If one electron or one million of them are to be considered in the equation, then their size, or the overall size of the one million, must take into account the size of the orbits of the neutrinos, since the difference in size can be significant, and gravity is calculated from surface to surface of objects.

There would be a big difference in size between one million electrons within the atom, where all the orbits of the neutrinos are shrunk, and one million electrons outside the atom, where all their orbits are enlarged. The number of electrons alone is not useful to our calculations to measure their speed, and neither is their inner expansion rate, which never changes, and is the same as everything else in the universe.

Larger clusters of electrons, such as heat, should travel faster than smaller clusters of electrons, such as x-rays. The same for light of longer wavelength or lower frequency, such as infrared, compared with ultraviolet light, which is of higher frequency or shorter wavelength. Infrared light should travel faster, although probably not by much. There is also the possibility that larger clusters will encounter more drag or resistance as they travel, reducing their acceleration. So, possibly it all equalises in the end.

I must admit that it is possible that, whether the clusters are larger or smaller, if the total number of electrons within these clusters of light remain the same, then overall, the size of the subatomic object will always be the same, whether the clusters composing the object are larger or smaller. And then, indeed there would be no difference in speed between infrared and ultraviolet light. Which overall would make it easier to measure the speed of any light or phenomenon in physics, composed of subatomic particles, such as radio waves.

This said, how did Mark McCutcheon reached the conclusion that the number of electrons overall would be the same no matter the size of the clusters? Is it possible he reached that conclusion in an attempt to explain why the speed of light is constant? Because if it is so, I just basically shown that the speed of light is anything but constant. Therefore, I feel the size of the clusters must make a difference in measuring the speed of light, that there would not be the same number of electrons at different frequencies, and that different types of light, travel at different speeds.

The Atomic and Subatomic Expansion Equation considers the size of the overall objects, when calculating the changing distance between these objects, instead of the size of its individual components. The larger the object made of clusters of electrons, no matter their individual size, the faster it should travel, because the distance decrease will accelerate faster the larger the object. The question is, is the overall size of the beam made of the electron clusters, the same no matter the light frequency, meaning no matter the size of the individual clusters composing the beam? Careful testing would be required to confirm this, one way or the other. Plus, the overall size of a beam of light, might not depend at all, on the frequency of light or the size of the clusters within it.

On top of that, there is something else to consider. We need to assess if, at the origin, the speed at which these clusters are launched freely into space, differs depending on their size. In order to free smaller or larger clusters of electrons from the subatomic realm, or from the atoms, differing amounts of energy might be required, more or less agitation of the neutrinos, electrons, atoms and molecules. Possibly smaller clusters are freed at a different speed than larger ones. Does it all equalise out in the end?

If smaller clusters of electrons are launched into space faster than larger clusters, I'm just assuming here, then they might travel and arrive faster. It is also possible that this added velocity of the smaller clusters is not that big, and possibly they would all arrive more or less at the same time. However, the smaller clusters' initial speed would be faster than the larger clusters, while the larger clusters would accelerate faster towards the destination, because of their larger size.

Plus, we need to take into consideration the overall size of the heat, the light or the radiation beam, which may very well be, as McCutcheon believes, the same size for all frequencies of light, no matter the size of the individual clusters composing the beams. And maybe they are all launched out of the atoms at the same speed. These are things we need to take into consideration when attempting to measure the speed of light accurately, depending on the different light or radiation frequencies.

When applying the Atomic and Subatomic Expansion Equation, each second we would have to re-enter the value of the distance between the objects, in order to calculate the changing distance, or gravity, at that particular moment in time. The force of gravity on the clusters will be different whether these clusters are near the source, for example the Sun, or virtually just above Earth. Although, one measurement of gravity at the beginning, from the source to Earth, should give us an overall good indication, about what the impact of gravity will have on the speed of light for that particular measurement.

The motion or speed towards us or away from us, of the source of the light, and of the electron clusters themselves, after they have left the source, must be taken into consideration when calculating the changing distance between the objects. Because that distance can change rapidly, due to the high speed of the clusters. We have to ensure that it does not play with our measurement of gravity. It is not only the size of the objects, and their expansion rate, that matters when calculating gravity, it is also their speed and their direction of motion, which affect each second the original distance between the objects in the equation.

The force of gravity on a bullet simply dropped from a plane, or fired at high speed directly at the ground, or fired at high speed towards the sky, where it will first go up and then down, will make a big difference in the bullet's velocity and how long it will take to reach the ground, even if the force of gravity remains the same in all cases.

Gravity in Atomic Expansion Theory, is no longer a force like it is in Newton. It is now a changing distance between objects undergoing an inner expansion, which causes the speed of light to accelerate. But if these objects are moving at high speed, in all directions, our measurement of gravity, and the impact it might have on the speed of light, could change as time goes by. I thought it was worth mentioning, so anyone attempting to calculate accurately the speed of light, will take everything into consideration, and ensure that the measurement of gravity and its impact are correct at all times.

There are other occurrences varying the speed of light which we must address, such as what happens when electron clusters in light go through glass, fluids, and other gases in space. Why the clusters, after they slowed down within the glass, while hitting molecules and atoms, resume at the speed of light once they exit the glass.

In the crossover effect, once in the glass, the neutrinos' orbits within the electrons have returned to their subatomic definition, orbiting more slowly in radically shrinking orbits. However, these clusters entered the glass with some momentum and speed, and some of them will go on to hit and destroy several atoms, which could then launch out more electron clusters at high speed. Many of the original clusters won't come out of the glass, being trapped within the molecules and atoms of the glass, or escaping as trapped heat instead within the glass, meaning larger clusters of electrons than light.

But the clusters that do escape the glass as light, are once again launched freely into space, and this happens when the neutrinos and electrons are orbiting faster in larger orbits. And so, it is expected that they will immediately continue on their way at high speed, once their orbits are fully extended and the neutrinos and electrons are free to move faster, even if perhaps they might go a bit slower than before. This would also apply to when they go through molecules of gas in space, or clouds of electrons in magnetic fields, whatever the electron clusters within light might collide with while travelling in space.

It is also likely that the original high speed of the neutrinos and electrons within the clusters of light, despite being prevented from moving freely at that speed while in the glass, don't really lose their initial state and momentum, they simply push back within the glass until they are out and can resume their original momentum, albeit perhaps a bit more slowly than before. After all, those are the particles that have not hit anything, and they entered with some momentum and speed already.

And, we may have the new clusters due to collisions with atoms within the glass, also adding to the ones that were lost. It is therefore no surprise that light should resume at high speed once outside the glass, and probably at the same intensity as before, or not much less. This said, it is very likely that any beam of light going through glass, or gases, or even magnetic fields in space, dissipates and reduces in speed as a consequence, as its clusters collide with atoms, molecules and electrons they encounter in space.

So, the overall size of the beam of light, might not be the same before and after going across matter in space. That too would affect the speed of light. Hence, we need to assess how much matter light goes through in space, to get an accurate picture of its speed. Light crossing a solar system full of magnetic fields, clouds of neutrinos, gases and other bits of matter, will not travel as fast as light between solar systems, where there is less matter. The drag or resistance might even stop the acceleration altogether here and there, like for falling objects on Earth, that stop accelerating when they reach terminal velocity.

If you hold and simply drop a bullet from a plane, like if it was one cluster of electrons, it will first have zero acceleration, free floating into space. But as the bullet and the Earth expand, the changing distance between them will accelerate, when we consider the two distance decreases when calculating gravity. The higher the bullet is dropped from, the faster it will reach the ground, because the more the decrease in distance will accelerate. Until it reaches terminal velocity, that is, where it would stop accelerating.

It may seem counterintuitive to imagine that the speed of light could be accelerating in the same manner, however, how could clusters of electrons not also be affected by the expansion of the Earth, like bullets? If a parachutist is dropped from a plane, and the parachute fails to open, that person is accelerating towards the ground, until he or she reaches terminal velocity and stops accelerating, due to air resistance. It does not seem counterintuitive then, so why should it be for light, if light is also made of matter. It would be a different discussion if light was still some sort of ethereal energy, like photons, but those don't exist in Atomic Expansion Theory.

The difference with clusters of electrons, is that they would not be light, heat or any other form of radiation or electromagnetic signal, freely moving into space, if they had not initially been going at a certain speed, as this is required to free themselves from the subatomic realm. As such, those clusters of electrons are more akin to these bullets being shot out of a gun from a plane, pointing downward towards the Earth. They would already be moving at high speed from this initial release from the subatomic realm, as they can only be freed if atoms collapse or explode. And possibly, they too eventually would stop accelerating due to the air resistance. It is hard to tell, since subatomic particles are much smaller than molecules of air, and yet, just like in the block of glass, some of these subatomic particles won't reach the ground, but they may create more small atomic explosions, freeing new clusters all the time.

This is why the clusters of electrons within light, will arrive much faster on Earth than the bullet simply dropped from the plane, but all will still have to obey the laws of gravity. Gravity due to the inner expansion of the objects, or the changing distance between the bullet fired from the gun and the Earth, must take into consideration the original speed at which the bullet left the gun. Just like for the clusters of electrons' initial speed, at which they were freed from the subatomic realm, needs to be considered when measuring how fast they will reach Earth, when adding gravity due to the inner expansion of the clusters and the Earth.

To summarise this section, the inner expansion rate of the electrons and the electron clusters, is the same as the one of the atoms, they expand at the same rate as the Earth. However, because the Earth is very large compared to the clusters of electrons, it expands much more proportionally. Neither is the enlargement of the orbits of the neutrinos within and between the electrons, in the crossover effect, or the high speed of these neutrinos and electrons, cause the electrons or clusters to expand faster. They can grow and shrink in size, but overall, at the base, their inner expansion rate is constant.

The required speed to be launched into space could depend on the frequency of light, meaning the size of the electron clusters within the light. The speed of light could also be affected by the size of the clusters, as larger ones should travel faster than smaller ones in gravity, but not if the beam they compose is overall the same size. Light could also be slowed down as it encounters a lot of stuff in space, like gases and magnetic fields, just like when it goes through a block of glass or water. And, light might be accelerated when passing by large objects, such as the Sun, due to the Slingshot Effect of partial orbits.

Both gravity and orbits are a consequence of the geometry of expansion, and I cannot see how light could be immune from its effects. And observing the behaviour of light, while travelling in space, certainly offers strong indications that light is being affected by gravity and the geometry of expansion.

These are all factors that can influence the speed of light, and most of them would influence the speed of light even in a vacuum. Only the slowing down of light due to matter in space, would not affect the speed of light in a vacuum, because then there would not be any matter there to slow it down. In its travel, light can vary in speed a great deal, decelerating and accelerating depending on the circumstances.

5.4 Gravity versus attracting and repelling forces

In chapter 3, I reviewed what is gravity within Expansion Theory, and the Atomic Expansion Equation, which has now replaced Newton's Law of Universal Gravitation to calculate gravity. I explained that gravity is no longer Newton's attracting force acting at a distance, or Einstein's geometry of space being a distorted rubber sheet of spacetime, where a larger object's imprint will attract smaller ones, like water going down a sink. Instead, gravity is now simply the changing distance between two expanding objects.

Even today in our physics we still use Newton's equations, because Einstein's gravity from General Relativity only makes a difference at speeds closer to the speed of light. Newton's main gravity equation, which has served us well for a very long time, shows something quite interesting: it is similar to the main equation of Coulomb's Law, which governs the branch of physics called electrostatics, or static electricity. Today this oddity in our science seems to be considered a simple curiosity. However, sometimes Newton is used instead of Coulomb, and it seems to work. Also, this issue is important to the proponents of the Electric Universe Theory, which is why I am spending some time on it.

Newton's equation calculates gravity between two atomic objects, while Coulomb's equation could actually be calculating gravity between two subatomic objects, meaning between objects made of electrons instead of atoms. This was my first thought, until I visualised this further.

Initially I thought there were two types of gravity in our universe, or two types of changing distance: the one between expanding atomic objects, and the one between expanding subatomic objects. This difference simply occurred, I thought, from the fact that both the atom and the electron have a different expansion rate in Mark McCutcheon's book *The Final Theory*. An atom expands much slower than an electron, explaining the difference between Newton's Gravitational Constant G, and Coulomb's constant K.

However, I have since assessed that the expansion rate of the electron is the same as the one of the atom. The difference lays simply in the initial speed boost gained by electrons, electron clusters and clouds, or compressed bands of electrons (radio signals and microwaves), when they are launched to freely expand into space. To that initial speed we need to add the gravity due to the inner expansion of everything else. Hence, there is just one type of gravity, and it affects atomic and subatomic objects the same way, no matter the scale.

Essentially, the speed of light is not constant in all frames of reference. Instead, the changing distance between the expanding clusters of electrons which light is composed of, and any observer made of atoms like a human being, could be measured as simple gravity between such objects: a changing distance between expanding objects, whether they are made of atoms or electrons.

As for Coulomb's Law being used to measure gravity, after some further visualisations, comparing the very small to the very large, it showed me that Coulomb's Law is linked to gravity, while not being gravity. It is still linked to the geometry of expansion.

Just think of a static electric field of two clouds of expanding electrons, meeting halfway between two positively charged rods of glass, as discussed in chapter 3. The glass rods are being repelled by their inner expanding clouds of electrons, pushing against each other, while the whole apparatus remains immobile, and could do so indefinitely if left untouched. The electrons remain in place by this exchange of neutrinos orbiting between them, and none of their orbits are enlarging or contracting. The electrons' growth has stopped, despite their ongoing inner expansion.

Compare this to the very large. This would be the equivalent of two expanding clouds of solar systems, all equalised to the same size, all being kept apart at a same distance, by their planets orbiting between each other and bonding them together. It would be quite a sight, if we could see something like this in the night sky. Unfortunately, almost all of the stars we see, are the ones within our own galaxy, and there are no static electric fields within atoms.

Gravity reduces the distance between objects due to the inner expansion of matter, an effect of the geometry of expansion, which is also responsible for the natural orbit effect. Those orbits are what keep these objects apart, by counteracting the distance reduction due to gravity. Newton measures the distance reduction between these objects, while calling it a force of attraction, while Coulomb measures the energy required, or force, for these orbits to keep these objects apart, counteracting the distance reduction.

Gravity does not disappear in any case, distance will always reduce between everything, because behind the scenes everything will never stop expanding. Consequently, there is a force required to push away these particles and objects, that the distance reduction is always trying to bring back together. There are no forces per se involved here, it is all simply due to the geometry of expansion and the natural orbit effect discussed in chapter 3.

To conclude, it is all related to gravity, but they are not the same phenomenon. And this is why in Newton the force is only attracting, while in Coulomb the force can both attract and repel, as those orbits can either expand or contract, depending on the geometry of expansion, which results from everything contained in that system.

Add more matter, and everything will orbit faster and the orbits will enlarge, causing a repelling force. Take out some matter from the system, and everything will slow down and shrink, causing an attracting force. There you have it in a nutshell, the difference between Newton and Coulomb. And now, let's compare the three equations.

5.5 Comparing Atomic and Subatomic Expansion Equation, Newton's Law of Universal Gravitation and Coulomb's Law

5.5a Atomic and Subatomic Expansion Equation, to calculate gravity between expanding objects made of atoms and/or electrons

$$\mathbf{D' = D - n^2 X_A \, x \, (R_1 + R_2) \, / \, (1 + n^2 X_A)}$$

D' is the changing distance between two expanding objects of radius $\mathbf{R_1}$ and $\mathbf{R_2}$

D is the original distance between the two objects

n is the number of seconds that have passed since the original distance was measured between the two objects

$\mathbf{X_A}$ is the universal atomic and subatomic expansion rate's constant (0.00000077 $/s^2$ or 7.7×10^{-7} $/s^2$)

The top of the second part of the equation refers to the absolute decrease in distance due to the internal expansion of the objects, and the bottom part refers to the further relative decrease in distance since space has not expanded along with the objects. This Atomic and Subatomic Expansion Equation calculates gravity at any scale, and replaces Newton's Law of Universal Gravitation.

5.5b Newton's Law of Universal Gravitation

$$\mathbf{F = G \times m_1\, m_2 \,/\, r^2}$$

F is the attracting force between two atomic objects of mass $\mathbf{m_1}$ and $\mathbf{m_2}$
r is the distance between the centre of the masses
G is the gravitational constant ($6.674{\times}10^{-11}\ N \cdot (m/kg)^2$)

It so happens that this equation is very similar to Coulomb's Law in static electricity.

5.5c Coulomb's Law

$$\mathbf{F = k_e \times q_1\, q_2 \,/\, r^2}$$

F is the attracting or repelling force between two charges $\mathbf{q_1}$ and $\mathbf{q_2}$
r is distance between the charges
$\mathbf{k_e}$ is the Coulomb's constant ($8.9875517873681764{\times}10^9\ N\ m^2\ C^{-2}$)

With Coulomb, instead of using the mass of the atomic objects, we are using the charge of the subatomic objects, which is a multiple of the elementary charge, usually denoted as *e* or sometimes q. The elementary or smallest charge possible, is the electric charge carried by a single proton, or equivalently, the magnitude of the electric charge carried by a single electron, which has the charge $-e$.

A charge must necessarily refer to how many electrons there are in a subatomic particle, which then must refer to the size of the subatomic particle. These three equations above, McCutcheon's, Newton's and Coulomb's, will work and be reasonable approximations no matter if we use mass, charge or size of the objects, because they only measure equivalence between certain variables and values.

However, Coulomb's Law does not measure the changing distance between expanding subatomic particles, and is not equivalent to measuring gravity, or the changing distance between two expanding atomic mass. If we could see and measure better at that scale, we would see that the Atomic and Subatomic Expansion Equation can replace Newton's Law of Universal Gravitation, but it cannot replace Coulomb's Law.

Essentially, the same gravity or changing distance between expanding objects, applies to both objects made of atoms and objects made of electrons. And although the Atomic and Subatomic Expansion Equation can calculate gravity between expanding subatomic objects, most of that changing distance between subatomic objects is counteracted by the orbits of the electrons within and between atoms, and the orbits of the neutrinos within and between electrons, which keep all these particles apart at a distance.

These orbits will grow and shrink depending on how many other particles there are nearby, how near these particles are, and the number of orbiting electrons and neutrinos around and between them all. They are seeking a balance in size and numbers, which is due to the natural orbit effect and the geometry of expansion. This is what defines charges in Atomic Expansion Theory, this ability of atoms and electrons to bond, attract or repel other particles, in a balancing act of smaller orbiting particles keeping them apart.

This attracting and repelling force due to the orbits of the neutrinos between the electrons, counteracts the distance decrease between electrons as they expand. Coulomb's Law mostly only applies to static electric fields, meaning in situations where everything is more or less immobile, and the orbits are neither enlarging or contracting. Therefore, that repelling force of Coulomb, is equivalent to the force of attraction of gravity in most situation, to prevent everything from crashing together, by keeping everything apart.

It is not the electrons themselves that expand in the crossover effect, it is instead solely the orbits of the neutrinos that will cause the growth or shrinkage of the electrons, and this is independent of the constant inner expansion of the electrons and all the particles within them.

Although, as a result of the inner expansion of the electrons, the orbits of the neutrinos are constantly enlarging slightly, and orbiting a bit faster. Just like our solar system is constantly expanding, as all the orbits of the planets are enlarging to compensate for the expansion of the Sun and the planets. Then, a static field is not completely immobile or static, even if we cannot see it or measure it, because we are also expanding.

What causes the growth and shrinkage of these orbits is the speed of the neutrinos. The faster the neutrinos orbit within the electrons and between them, the larger the orbits and the growth of the electrons, and the more the force of repulsion will be. The same applies to solar systems with planets orbiting between them, and galaxies with solar systems orbiting between them.

5.6 Gravity is the changing distance between expanding particles or objects, while attracting and repelling forces are due to the changing speed and orbit's size of smaller particles or objects, orbiting between larger ones, counteracting gravity

Orbits in Atomic Expansion Theory are calculated using Kepler's Three Laws of Planetary Motion, and a new equation Mark

McCutcheon has called the Geometric Orbit Equation. McCutcheon demonstrates in chapter 1 of his book *The Final Theory,* how this new purely geometric equation was previously unrecognised, and is readily extracted from patterns in the standard astronomical data. It shows that the orbital radius R of any planet in our solar system (i.e., its distance from the Sun) times the square of its velocity v, always gives the same constant value K, which gives the new equation $v^2R = K$.

McCutcheon also shows in a derivation, how this purely geometric equation, has been incorrectly embedded by Newton in his gravitational and orbital equations, unnecessarily adding mass, in calculations of orbits when there was no need to. No doubt this is responsible for many required course corrections in our orbital calculations today.

There is gravity for objects in orbit, or a changing distance between orbiting expanding objects, however it is counteracted by the natural orbit effect of the geometry of expansion. When behind the scenes, a planet is moving away from the Sun in a straight line, fast enough to escape the distance decrease due to the expansion of both the planet and the Sun, then in our resulting reality, where we can no longer see the expansion, the planet enters an orbit around the Sun, and the changing distance between these two objects, keeps getting larger and smaller alternatively, sometimes forever if left untouched.

Behind the scenes, from the first perspective, they are still both expanding and moving away from each other, most likely in straight lines. Depending on their original speed and direction of motion, which never change unless they hit something, the distance between them will keep enlarging, will remain the same, or will reduce until they collide.

In our resulting reality, in the second perspective, there is still gravity in effect when objects are orbiting each other. The distance between the planet and the Sun would normally decrease until they crash into each other, if it was not for the planet moving fast enough to escape the expansion of the Sun. Even if the planet orbit seems constant to us, from the third perspective behind the scenes, its orbit constantly enlarges in a spiral. This is not exactly yet an attracting or repelling force, but it is the beginning of it.

Now imagine that the planet in question was orbiting around or between two suns in two different solar systems, and that this planet, along with other planets on different orbits between the two solar systems, were capable, through their orbits, to keep these solar systems apart, counteracting gravity.

If more planets were to come in from nowhere, suddenly the planets would orbit faster, their orbits would enlarge, and the distance between both solar systems would increase, just like a repelling force. If instead, some planets were destroyed, or moved away from the system, the geometry of expansion would slow down all the orbits of the planets, the orbits would shrink, and the distance between both solar systems would decrease, just like an attracting force.

The number of planets within each solar system, is what will define if the force will be attracting or repelling. The more smaller objects you have on one side, the larger that overall side is, the more gravity there is through the geometry of expansion, the more the speed of all the objects and the size of their orbits will be affected. Clouds of solar systems, just like clouds of electrons, are kept apart respectively by an exchange of planets and neutrinos, which change speed and the size of their orbits depending on how much matter there is in their surroundings, due to the geometry of expansion.

This is how electrons can attract or repel each other or other particles such as protons. My new definition of a proton is a very large cluster of electrons bonded together, or a very large cluster of solar systems bonded together at a higher scale.

With Coulomb, if the two charges q_1 and q_2 have the same sign positive or negative, the electrostatic force between them is repelling. If they have different signs, the force between them is attractive. And if you remember from the previous section, a charge is defined by the number of electrons on the surface of an object. The equivalent of these charges, at the larger scale, would be how many solar systems there are, while how many planets they have would define if the force will be attractive or repelling.

In Expansion Theory, *charged* particles can only have the structure of a nucleus with orbiting smaller particles. With motion, an exchange of the number of the smaller orbiting particles is what creates the bonds by which they can attract or repel other particles, and this causes the distance decrease or increase between them.

In gravity there is a distance reduction due to the inner expansion of the objects, and there is an additional unrelated attracting and repelling force due to the orbits of *smaller* particles or objects, linking together *larger* particles or objects. This additional attracting and repelling force, can counteract the first distance reduction due to inner expansion, by keeping the particles or objects apart.

In the previous chapter, I describe my new theory of the model of the electron, which essentially has a nucleus with orbiting smaller neutrinos, very much like the structure of a solar system. And the motion of an electron, simply follows the same natural orbit effect of the geometry of expansion, just like for solar systems in the very large.

In Atomic Expansion Theory, there are no charged particles, everything is explained by the simple expansion of matter. Bonding of particles like atoms to form molecules, is the result of an exchange of orbiting electrons between atoms. The same way, I explain how electrons can bond with other electrons and protons, by the exchange of neutrinos orbiting between their nuclei.

All objects that have a nucleus with orbiting particles, can attract and repel similar objects or groups of them. And the bonding of such particles, depends on the internal number of orbiting particles around the object, and between the objects.

In their search for a balanced system, entirely due to the geometry of expansion governing gravity and orbits, if objects have fewer orbiting particles than others, they will share these particles and attract each other. If they have the same number of orbiting particles between them, they will keep each other at bay at the same distance, as there is nothing to balance. If suddenly both objects gain a surplus of particles, everything will start to move faster in enlarging orbits, and they will repel. Not only this explains all molecular structures of different atoms gluing to each other to form the chemical elements of the periodic table, but it also explains the mechanics of all chemical reactions.

As I discussed in detail in the previous chapter, an electron, which is at the base of the electric charge, is the same as a solar system at a larger scale. The planets orbiting the Sun are essentially neutrinos orbiting the nucleus of the electron. Whatever the behaviour of electrons, it must be compared to the behaviour of solar systems. Just like the behaviour of galaxies must be compared to the behaviour of atoms.

Without orbiting particles, these electrons would be neutral particles. Planets are just neutrinos, they are charge-less particles, even though they too have orbiting objects around them, their moons. Technically neutrinos could also have a charge, attracting and repelling other neutrinos, bonding with them. For that, we would need to have moons orbiting between two planets, which we have never witnessed yet. As far as we know, so far, moons only orbit their own planets. Hence, planets and moons are considered charge-less objects, just like neutrinos. Some do have magnetic fields however.

Without a charge, meaning without an exchange of particles, planets cannot repel anything. Planets simply attract in their inner expansion, meaning the distance decreases between such objects, due to the internal expansion of all objects. Although such objects will still follow the natural orbit effect of the geometry of expansion, so they can still orbit other objects without being *charged.* Objects can only be charged, if they exchange smaller particles between them, by which they can attract or repel.

Solar systems on the other hand are *charged* particles. They have a nucleus, they have orbiting particles as their planets, and other orbiting particles or planets are orbiting far and wide. Our solar system probably has several more planets as yet unidentified, orbiting very far from here, up to the distance between our solar system and its neighbours. Far enough, that if two solar systems have an imbalance of planets, some planets at the edge, or near the middle of both solar systems, will start orbiting between these two solar systems, although we have never seen it yet, in our short history.

There are rogue planets between solar systems, we have identified a few, but we have failed so far to calculate their trajectories, and confirm if they are orbiting between solar systems. They just take a long time to orbit between solar systems, and they never come close to the Sun or our known planets. Our eight planets and the Sun, are just the nucleus of the solar system.

Depending on how many planets a solar system has outside the solar system, compared to another solar system, a balance will be reached, and they will either attract each other if they don't have the same number of planets, stay at the same distance if they have the same number of planets more or less the same size, or repel each other if they both have a surplus of planets, or if there is an incoming flow of planets within the entire system coming from elsewhere, like it might happen if other solar systems were to approach.

It is the same for galaxies, which are essentially atoms. The number of solar systems within them, and orbiting them well outside the visible galaxy itself, and orbiting between galaxies, will define if the galaxy will bond with other galaxies, through the exchange of solar systems, and reach a balance, or if they will repel each other.

Galaxies are composed of the equivalent of larger particles like neutrons and protons, which are simply several clusters of solar systems bonded together. The overall size of these clusters within galaxies or atoms, and the overall size of these galaxies or atoms, will affect gravity and the orbits of all these smaller particles. This is likely what defines if a balance must be reached or not, and if the galaxy or an atom will bond with another, and at which distance.

It is all a question of the size of all the different objects, which affects gravity and the speed and size of all the orbits, that we measure through *charges* and *forces* using Coulomb's Law. But overall, Coulomb is just a convenient equivalence, as it is all the geometry of expansion which dictates everything, and there may be better ways to measure all this, now that we understand what is truly going on.

Not only Atomic Expansion Theory explains Newton's attracting force, but it also explains Coulomb's attracting and repelling forces. As a bonus, it works at all scales, whether we are talking chemistry or astronomy. It is all geometry based, there is no need for forces, mass or charges in science. There is only need for size, changing distances due to expansion, which is gravity, and orbits' size and speed of all the particles and objects interlinked together, which is the natural orbit effect of the geometry of expansion.

And, in theory at least, if not in the practical sense depending on the scale, all of this can be calculated using the Atomic and Subatomic Expansion Equation, Kepler's laws of planetary motion, and McCutcheon's Geometric Orbit Equation. This is what a true Theory of Everything is meant to achieve.

An ion is an atom or a molecule in which the total number of electrons, is not equal to the total number of protons, giving the atom or molecule a net positive or negative electrical charge. So, for a galaxy, it would depend on the number of solar systems orbiting very far from the entire galaxy (the nucleus), compared with the number of clusters of linked solar systems within that galaxy (protons).

This way, we can see what Coulomb's Law is referring to at the larger scale, the forces required to keep apart solar systems and galaxies, in a universe where distances constantly reduce, due to the inner expansion of everything.

It is difficult to notice the changing distance between expanding electrons, or Newton's gravity at the subatomic scale, because electrons are seldom alone in their movement. The distance decrease is more often, if not always, counteracted by the orbits of neutrinos keeping them apart. The orbits of these neutrinos, or of the planets, and their numbers, is the additional attracting or repelling force counteracting the decreasing distance, that is due to simple inner expansion of all these particles.

However, the inner expansion of these particles still has an effect. The orbits of all these particles, as well as the entire objects formed by these particles, are constantly expanding as well. Although for us this expansion is unseen, as everything expands proportionally at the same rate.

And finally, it can be seen that solar systems and galaxies can be very strong objects, as strong as electrons and atoms, and can achieve the very same feat. They can be squeezed and they can push back, without crashing into each other, since there is an entire network of planets and solar systems being exchanged between them, as soon as they get near each other, keeping them apart, but also capable of reducing and increasing the distance between them. Although of course sometimes, they do crash into each other.

This was the last argument of Mark McCutcheon, to justify that atoms could not be galaxies, and electrons could not be solar systems, as solar systems and galaxies were fragile objects that could not do what electrons and atoms can do. And I have just shown that they can.

SIX

Our physical world is a psychophysical reality with no distance

6.1 Jane Roberts and other channelled material - Are we living in a simulated world?

In my opinion, a theory of everything really must cover everything. And if you have seen a UFO, doing impossible things in front of you, as I did when I was a kid, you might wish to explain it from the point of view of physics, or else you know you still don't have all the answers.

I have reviewed some of the physics contained in the channelled material from authors like Jane Roberts with *The Seth Material,* more specifically *The Early Sessions* book series 1 to 4, as this is all I had the time to read so far. I also pondered on several other sources, discussing physics related to the New Age, the spiritual movement and aliens, like *A Journey to the Multiverse* series of Wes Penre, the books of the author of *The Mothman Prophecies* John A. Keel, and the past lives regressions of Dolores Cannon, as found in *The Custodians* and *The Convoluted Universe* book series.

Our actual science is so far removed from anything they are saying, we could only just register everything as interesting concepts to be contemplated, but nothing more. For us to start considering New Age physics, a completely new theory of everything would first be required, one more akin to their vision of the universe.

There is also this concern of using terms we are familiar with, in order to help us understand difficult concepts, but not with the exact same definitions we use. Words used in channelled material, might mean something else than what we think. This is quite an important point, because often the channelled material appears to support actual concepts in our physics, seeming to validate our Standard Physics, when truly they mean something entirely different.

Most physicists and scientists today consider all channelled material to be New Age mumbo-jumbo. Several New Age authors, such as Deepak Chopra, have tried to marry the channelled material with our actual science, like for example the multiple dimensions of the String theories, or other bizarre quantum mechanical concepts. Atomic Expansion Theory replaces all these theories, and any other candidate for a theory of everything.

It requires quite a leap of faith to get into *New Age Physics.* The first one, that our reality could be psychological or psychophysical in nature, and consciousness the very source of our physical reality. Even that would not be the entire picture, as behind our psychological reality, could be an electrical universe, filled with electrical signals and impulses of different frequencies, codes our mind and brain would interpret to instantly construct our physical reality. Very much like our technology does today, but on an electrical, chemical and biological level. As a result, the real primary universe would simply be a blanket of expanding electrons, that New Age authors call the pure energy of the universe, and from it everything would bursts forth, through electrical signals of different vibrational frequencies.

Could this world have been created, either through a God or a computer programmer? Who can really say with any kind of certainty? I consider everything to be possible, even the existence of God, which Seth in Jane Roberts defines as a self-aware energy field, called All That Is, that slowly became aware of its own existence, and who, in thoughts, formed portion of itself into units of consciousness, creating and animating entire universes. Something we all do to a lesser extent, in our mind through our imagination, I suppose. It is my intention with this book that *New Age Physics* should bring religion, spirituality and science together. Wouldn't that be a true and worthy theory of everything?

If the philosophy Professor Nick Bostrom, from the University of Oxford, is correct, and we are living in a virtual reality - and he is presenting a compelling argument in his paper *Are you living in a computer simulation?* - then the idea of a creation event instead of an evolution becomes very likely. A creation which somehow contains an entire history of the evolution, to fool us into believing this world is immanent and has always existed, when in fact it might have been created this very morning by our own brain, when we woke up and opened our eyes.

This said, what we collectively re-create in our mind, comes from patterns and blueprints out there, according to Seth, and they are always there. And so, evolution might also be real, even if we have to re-create it in our brain, along with the world, like individual computers do, in order to experience it. At this time, we just don't know how it all works. Nothing has to be mutually exclusive, we may just have to find a way to sensibly explain it. We may all be talking about the same things, except we might be using different words, definitions and vocabularies, in order to define these concepts.

In a computer simulation there is no real distance to speak of, it is a virtual space where objects are either larger or smaller than others, and where to simulate motion, these objects simply expand or shrink in relation to each other. And there you have it, in a nutshell, my entire Universal Relativity Theory, which will be discussed in more detail, along with the implications, in the next chapters.

Please note that this theory does not require that we live in a computer simulation or a holographic universe to be true. It may simply be that, nature is constructed very much the same way that we can replicate it as a virtual reality, using computers and the same components found in nature, such as the electrons.

The electron is the only fundamental particle in nature, according to Atomic Expansion Theory, everything is made of electrons, and we are living in an electronic world. Although for me, the most fundamental particle in nature is the neutrino, and everything is made of neutrinos of varying sizes. Certainly, this theory about the expansion and contraction of matter, would still be true whether or not we were living in a computer simulation.

Is New Age truly just mumbo-jumbo? Or will it suddenly make sense once we discover a true theory of everything in physics? This is precisely what this book is about. We have a theory of everything, it is called Atomic Expansion Theory by Mark McCutcheon, along with my new Theory of Universal Relativity, which complements it. When we read the New Age literature in light of this theory, suddenly let there be light, as it appears compatible. You will see in the next chapters how a lot of the channelled material, can suddenly be brought to a real hard science that can be observed and measured.

This book is a work in progress, subject to change as I read more of the channelled material. This is the kind of work that would take a lifetime to research and write, and I didn't have that kind of time in my life. So let's see what I have uncovered with my limited time, and hopefully others will pick up on this, and together we will come ever closer to the truth about the world we live in.

This has always been the meaning of my existence, to identify what the universe is, to understand who we are as human beings, and our place and purpose within the universe. When we take a step back to look at the night sky, and consider the planet we live on, and then look through a microscope to see another world teeming with life, and further down we see these atomic and subatomic particles; one must admit how crazy this place is, as none of it ever made any sense to me. What in the world is this all about? If we reach the end of our existence a bit less ignorant, then maybe this life will have been worth it. If you are new to Atomic Expansion Theory, you are up for quite a discovery. It could radically change the world of science, and forever change how we perceive the world we live in.

> "Your own universe expands as an idea expands, in ways that have nothing to do with space."
>
> Seth, Session 149, Book 4 of *The Early Sessions*

> "Conceptions explode into being, evolve, change into something else, and yet all this within a framework that is definite but cannot be seen or touched by you, the psychological expansion even of atoms and molecules that in themselves contain condensed comprehensions beyond your understanding. The stuff of the universe is not inanimate, nor even on your plane is it an emptiness to be filled, but constantly expanding in ways that you cannot yet fathom."
>
> Seth, Session 43, Book 2 of *The Early Sessions*

A lot of the channelled material discusses science, including the physics of our reality. In fact, some of them seem to cover several planes or systems of existence (realities), or at the very least several dimensions or world densities. However, none of the channelled material appears to reveal a well-defined theory of everything in physics. For what is still our primary physical reality, it seems to rely on almost nothing from our physics. However, some authors have drawn parallels with Standard Physics, like Quantum Mechanics, and both Theories of Relativity, and even Newtonian gravity.

We know these theories are hopelessly incompatible, and that even Einstein's Relativity, taken on its own, is so incorrect, that in order to make it work we had to invent hypothetical dark matter, dark energy and some Big Bang, which are all unlikely concepts. I have written an article about this included in this book: *Dark-Matter, Dark-Energy and the Big-Bang All Finally Resolved.*

Seth even implied in Session 43 of the second book of *The Early Sessions,* that there was no Big Bang, and clearly stated that the universe is not expanding ever faster as defined by Inflation Theory. In that session, Seth described what he means by the expansion of the universe:

> "The universe is expanding in the way that a dream expands. In other words this expansion has nothing to do with <u>your</u> (underline) idea of space. The expansion, in a most basic manner, is more like the expansion of an idea. It has nothing to do with space or time in the manner in which you are accustomed to think of them. I told you earlier that your scientist's idea of an expanding universe was in error, although in one important sense the universe was expanding, and this is what I referred to."
>
> "Take two paintings of the same shape and size; that is, two paintings that take up the same amount of space in your universe. One painting is extremely crude and poorly done. The other painting not only seems of superior quality, but also appears to undergo a continual transformation while still taking up the same amount of space.
>
> "Say that the painting is a landscape, and that this transformation includes within the same amount of space the addition continually of more trees, more hills; that the hills still within the same amount of space allotted to the painting grow taller, and yet never shoot past the frame. Imagine this transformation including also the addition of distance, reaching <u>ever further backward</u> while not disturbing the <u>back side</u> of the canvas.
>
> "Imagine if you can the figures or inhabitants in the painting having psychological reality, all within the set limits prescribed by the given space. Imagine in other words consciousness, growth, reality and expansion, having nothing to do with expansion of

space in your terms, but an almost complete freedom of psychological realities, and you will come at least within the realm of understanding what I mean by an expanding universe that has nothing to do with the expanding universe of which your scientists speak.

"Most realities have absolutely nothing to do with space as you imagine it. Most realities have their growth and existence in something closely akin to what we have called psychological time, and this is completely independent of space as you conceive it to be. Psychological time is a sort of climate or environment conducive to the existence of all consciousness.

"You think of space as an emptiness to be filled because on your plane you fill what you call space with camouflage patterns. And I repeat: instead true space, fifth dimensional space, is the vitality and stuff of all existence itself, vital and alive, from which all other existences are woven through means which I have outlined so far in a rather sketchy fashion. Even on your plane quality, which represents a kind of expansion, does not necessarily imply an expansion of space. The universe expands continually in a qualitative manner that has nothing to do with space as it is usually envisioned. And the expansion is more vivid and valid than I can possibly explain to you at this time."

Seth, Session 43, Book 2 of *The Early Sessions*

The word *expansion* is one word that comes back again and again across all channelled material, no matter the source. It appears nearly at every page, and often more than once, in the book of Esther Hicks *Ask and It Is Given: Learning to Manifest Your Desires,* where she channels an entity called Abraham. By far, it makes the word *expansion,* Abraham's most used word. And I bet it is also Seth's most used word. Do you think they might have been trying to tell us something, by repeating over and over again that everything is expansion? They must have been wondering when we would finally get it. I even read that *stretching out the heavens* is a popular expression in the Bible, and that there are many references to an expanding universe.

Expansion is a key word used everywhere in almost any circumstance. Expansion is central to the definition of our reality. Only a handful of people in this world have all the knowledge to understand what a theory of everything in physics should answer, and if you have read some of the channelled material, you might be one of them. Let's review these most important points.

6.2 What a Theory of Everything needs to take into consideration

Seth, in the books of Jane Roberts, states that there is no distance in this universe, he repeats it numerous times in the first book of *The Early Sessions.* There is also no future, no past, no spacetime concept. There is just an instantaneous and simultaneous spacious or large present, where our entire history happens simultaneously. He says our history is fluid, meaning it changes constantly. The future as we conceptualise it, is constantly changing, but so is the past, and they influence each other. Humanity seems to follow a fluctuating timeline, where moment by moment, anything can change.

That alone seems quite an insurmountable hurdle for anyone working in Standard Physics. Our entire view of physics would need to radically change, before we can even start to conceptualise and theorise a universe without distance or time, let alone without the four-dimensional spacetime, where the past, present and future are constantly fluctuating.

Then Seth says that all is expansion, that our physical reality expands just like an idea in our mind would expand. He also clearly says that the atoms and molecules expand, and it is a psychological expansion, hence our universe is more akin to a mental construction. He goes further, the dream world and the world of thoughts are as real as our physical reality, that they too are made of electrons, atoms and molecules. While they are not experienced via our five outer senses, instead, they are experienced through our inner senses. There lays the reason we might feel it is not as real, although these imagined realities are as real as our physical one, and are made of the same particle physics. He also declares that this life energy of All That Is, the vitality of the universe, which is the basic unique thing that truly exists, and from which everything is made of, is matter.

And by saying that energy is matter, it is not the same thing as saying energy and matter are equivalent and interchangeable into one another, as Einstein states with his equation $E=mc^2$. According to Atomic Expansion Theory, energy is expanding matter. There is no ethereal energy, there is only expanding matter in clouds and clusters of expanding electrons, as a driving force to explain electricity, magnetism, electromagnetism, light, heat, and so forth, as shown in chapter 3. This said, moving matter, including light, even if moved through relative motion, possesses kinetic energy, and can have a real force of impact on objects. In that sense, Mark McCutcheon has called kinetic energy, the only form of energy that can be considered as such.

Energy made of matter in Atomic Expansion Theory, means that this life energy called All That Is, is made of electrons. This means that every single layer of reality, all these other planes of existence, 4th density, 5th density, 6th density, they are all physical realities made of matter. They are just of a different density of matter. Density here means their particles are larger or smaller, and more or less spaced out, when compared to our particles in our reality.

And these realities or universes, would also be of different dimensions, meaning of different sizes, and would involve the frequency at which matter vibrates. The faster electrons are orbiting around and between atoms, and the faster atoms are vibrating within molecules, would define a different dimension or world density, most of which we could not see, as we are limited in what we can perceive by the vibrational frequency of the matter composing us.

This is precisely like how we only see a certain frequency of light from the electromagnetic spectrum, the small section we call visible light. We don't see infrared and ultraviolet light, or x-rays and gamma rays. That there could be magnetic fields made of electrons vibrating faster, and matter made of atoms and molecules vibrating faster outside our range of vision, is not so farfetched, even from the point of view of Standard Physics. That there could be worlds out there inhabited by entities vibrating faster, including spirits and what the New Age calls beings of light, is quite possible and would explain a lot of the religious, spiritual and paranormal phenomena.

Then Seth tells us more or less that this entire reality is psychological in nature, what we think is what we create. By the time you reach book three and four of *The Early Sessions,* there is an entire electrical reality superimposed over the physical reality, it is made solely of energy (electrons), with electrical impulses of different frequencies which amount to codes that our brain transmits, receives, interprets and translates into some reality, be it physical, mental or dreams.

To make things even more complicated, at every moment reality splits and expands in all directions. Several physical bubble realities are being created, as many as there are people or anything conscious and alive, on the planet and in the universe, and as many bubble realities as each one of them can create in their mind, subconsciously or consciously. And they continue to exist and evolve on their own, even after we have left them and moved on to other thoughts and dreams. In these bubble realities that we each create, more alike several dreams we might construct every night, only us is real, everyone else is our own personal virtual construction or an avatar.

We all know what to build collectively, because we are all interconnected telepathically through these instant electrical signals in the electrical layer of our reality, so we all relatively build the same individual reality. These signals are faster than the speed of light, they are not part of our camouflage reality, since our physical reality is made out from the energy of this electrical layer, so it must primarily exist outside of it.

We don't always build the same collective reality. People with mental disorders, or under hypnosis, or under the influence of hard drugs, will construct different realities. This is not a question of perceiving reality differently, *they are* different constructs of reality, since all physical realities are mental constructions to begin with.

There are patterns or blueprints out there perceived by all of us, into which we project matter into, in order to build our personal reality. No pattern ever disappears, so even the past can be re-created at will, and probably the future, hence the idea that there is no time in the universe, just a spacious present. Our consciousness, our mind and thoughts, create new patterns through electrical impulses, and our brain constantly replaces all the matter in our own physical world at certain frequencies.

Our brain continually replaces the matter (electrons and atoms) of our reality, just like a computer would use the electrons within the electricity from the main, to constantly refresh a screen at a frequency of 50 or 60 Hz, which is based on the frequency at which the matter composing human beings usually vibrate. For the brief moments on our monitors when the display flickers, there is nothing to be seen on the screen, although this flickering is too fast for our eyes to see. This would then be a screen made of antimatter (absence of matter), instead of matter (electrons), showing an image.

Our universe would always exist as much as a universe of matter as a universe without any matter, while matter is constantly being replenished or replaced. This universe made of non-matter, or of the absence of matter, is what Seth calls a universe of antimatter or negative matter. This definition is far removed from the one for antimatter in our physics.

According to Seth, there is life and entities living in this universe made of antimatter, because during the time that there is no matter for us in our universe, there is matter in their universe, and vice versa. They too cannot see the constant absence of matter in their universe, as matter is being replaced too quickly for them to notice. And it may be due to what Seth calls a before and after image of our universe, as the matter is just about to disappear, and just about to be replaced.

There are no such patterns or blueprints to hold matter in our thought or dream realities, because time is different, it is psychological time and there is no need for a more permanent type of matter. Our thoughts and dreams are *individual* realities, compared to our awaken *collective* reality that everyone must build, more or less the same way, following those patterns. Seth says much more than that, but for now this will suffice.

Thus, someone will need to identify a theory of everything in physics that can explain all this, starting with the astonishing fact that this is a psychological or mental reality, built from our consciousness, that there is no distance in this universe, there is no space, no time, no four-dimensional spacetime, and yet, can we deny that this physical reality made of solid matter, exists and appears to be the same for everyone?

6.3 Fourth type of expansion according to Seth/Jane Roberts, concerning the nature of our physical reality

Seth states that our physical reality, or all the matter (energy) in the universe, expands from one point within our brain and is projected out to create the universe. This reality is psychological in nature, we create it in our mind, from our consciousness, and our brain re-creates it using energy from the electrical layer of our reality. So it must first be created from one small point called a mental enclosure, and then projected and expanded into a tapestry around us, of everything we can detect with our five outer senses. This he calls the camouflage reality, it is not real, although it certainly seems real to us.

Dreams are apparently closer to the real reality of which we are still unaware, since in dreams, time and space are kind of deconstructed or nonlinear, and thoughts can more easily create anything we want. Achieving this deconstruction is more difficult in a collective and shared physical reality, especially if matter vibrates more slowly and is denser as a result.

This psychological reality we would project out of our brain, possibly from the pineal gland, represents a rapid expansion of the physical universe, an instant expansion of electrons, atoms and molecules. It is the expansion of our reality, just like the expansion of an idea created within our mind. Because reality, although physical in nature, is still just an interpretation of our consciousness. It is electrical impulses and signals of different frequencies, that our brain interprets to create the reality that we see and feel, like a computer does.

According to Wikipedia, "the pineal gland represents a kind of atrophied photoreceptor. In the epithalamus of some species of amphibians and reptiles, it is linked to a vestigial organ, known as the parietal eye which is also called the third eye". This is not a mystical third eye, it is a real atrophied eye that used to see, and perhaps still can, and possibly it is the way by which we see our dreams and what we imagine in our mind, who really knows?

Wikipedia also states that "the pineal gland produces melatonin, a serotonin derived hormone which modulates sleep patterns in both circadian and seasonal cycles". The centre of our brain that Seth talks about as the mental enclosure - from which any reality we create and perceive, be it our imagination, dreams or our physical reality - could start with the pineal gland. Or some of these realities could be perceived through it. This is not something Seth has specified, he talks instead of several internal senses that we would have.

We each create our own bubble realities. Each time we think something different, a new one is created and starts expanding. We all relatively create the same realities through an instant telepathic connection of electrical impulses, signals and codes of different frequencies. Hence, reality is equivalent to a dream or imagination or thoughts. There are different realities or timelines of different frequencies, and you move to these different timelines through thoughts and will alone, as thoughts, feelings and emotions change your vibrational frequency.

This fourth type of expansion must be exceedingly fast, because when each of us open our eyes every morning, or whenever we walk to another room, we re-create a new reality. So almost instantly we project out of our brain this reality right into the furthest distance out in the sky, to the furthest galaxy we can see. And once it is there, matter continues to expand but at a much lower rate, and then the first type of expansion discussed earlier takes over.

From there on, physical reality - and it is the same principle for any other reality or plane or system of existence, that we individually create or collectively share - is made of real expanding matter. And this is the physical reality science has been trying to explain, and for which we are attempting to find a theory of everything.

6.4 The computer network analogy to explain the Many-worlds interpretation of our reality

This is not as esoteric as people might think at first sight. It is very much how a computer would build and expand a virtual reality from one single small point, using the electrons from the main (electricity), to be shared by millions over a network, following the exchange of electrical impulses of different frequencies.

In a way, there are as many bubble realities as there are computers and consoles connected to the network, and as many different realities as there are versions of the software constantly changing, as people accomplish or change things in the programme. The difference here, in the world of electronics, is that a computer only deals with electrons, it does not build atoms and molecules. Or not yet anyway.

This analogy with computers and networks, to describe our collective reality, would not have existed in 1963, when Jane Roberts started channelling Seth, and yet it is clearly stated in *The Early Sessions.* I must stress that this is just an analogy, although it is not surprising that our technology would reflect how our reality truly works. It must make use of what exists, like electrons, electrical impulses and signals, and electromagnetism.

Logically, if you just pop instantly an entire physical reality out of your brain, and can change it at will, then there cannot be any real distance or real space out there. Just like time cannot have any meaning, if all realities are solely psychological, and if you can just focus and imagine yourself in a different timeframe on a different timeline (bubble reality), and suddenly wake up there by re-creating that reality with and within your mind. Several psychic mediums have achieved such feats.

We can all imagine and create anything at any time, and make it come true independently from time and space, or from the collective memory of what reality should look like to everyone. This is probably our biggest obstacle, to free ourselves from the collective reality of everyone else around us, constantly telling us how our bubble reality should look like. However, if individually we focus hard enough, we can re-create any physical reality we wish. I have experienced this for myself, I am writing a book about it called *Changing Your Future,* but it is unfinished. If you search online, you can find early versions.

I must point out that this multiverse we would create that Seth describes, is very much in line with the Many-Worlds Interpretation of Quantum Mechanics. And it is re-iterated by the people Dolores Cannon speaks to under hypnosis in her books *The Custodians* and *The Convoluted Universe* series. Within Atomic Expansion Theory, Quantum Mechanics is no longer required, as Mark McCutcheon has shown that all these weird phenomena can easily be explained by the expansion of matter. As a result, there is something very similar to the Many-Worlds in Jane Roberts, explained instead by the fact that we can think any number of realities in our mind, and re-create them, just like a computer can run an infinite number of times the same simulation, but slightly different each time.

Just like a computer computes in its CPU all the possible moves that could happen in a game of chess before it makes its move, all these possible moves exist for real, or as real as the reality of the game, just through the simple fact that the computer has thought them all before playing. Thinking is akin to creating realities, so as soon as we start visualising all possible probabilities of an event, we create all these probabilities in our mind, and so they all have an actual reality made of electrons, atoms and molecules.

We just don't see the matter of all these multiple realities, because it would be vibrating faster. But there are a number of people who can see these thoughts as realities, as it is described in the book *The Holographic Universe* by Michael Talbot. They see our thoughts and our history floating around us. It is most probably how psychic mediums can see so much, and yet be often incorrect, since they see probabilities of a changing future, and even of a changing past. Psychic mediums would simply be people made of matter vibrating faster than normal, giving them the chance to see more.

There is no reason why we should not be able to devise technology that can see matter and realities vibrating faster, and suddenly record very clearly our thoughts and dreams. We need to devise technology based on the vibrational rate of matter, and that way we might even be able to see the probabilities of our future and of our past, our fluctuating timeline, who knows.

But just like every move the computer would have considered in the game of chess, that could actually represent thousands of possible configurations of the game, only one move will actually be played in our reality. And this is what we will perceive as real, even if all these possible realities will go on to exist as soon as they have been thought or computed.

If all realities are simple electrical impulses and signals, then any electrical impulse and signal of any kind and of any frequency, including what we think and compute, must have a reality floating around us. This is what quantum computers may be trying to exploit, but they would first need to understand the nature of quantum entanglement from the point of view of Atomic Expansion Theory, which I describe later in an article called: *Breakthrough in Faster-Than-Light Travel and Communication, and the Search for Extraterrestrial Intelligence (SETI)*.

6.5 Our physical reality appears to be psychological in nature

In the permanent physical reality, we have matter that appeared out of we don't know where, that expands for no reason that we can explain. If our physical reality is instead psychological in nature, and is built from one small point at the centre of our brain, where energy comes out, and expands rapidly to fill the tapestry of our bubble reality, then we know where it comes from, and we know why matter continues to expand once it has fully expanded into the reality we see.

It may be that there is a whole physical universe out there, with large distances, that has existed for billions of years, whether or not there ever were anything alive in that universe who could sense and detect it.

However, it is also possible that this universe never truly existed physically in the first place, except in some sort of archetype, or model, or blueprint, with patterns holding matter into place, shared by every living thing all communicating telepathically in a wireless network. We would then build part of it, each our own bubble reality, on a mental or psychological level, right from the centre of our brain. We need to review the arguments for the likelihood of such a psychological physical universe.

6.6 No difference between a psychological reality and a real permanent physical one

It makes no difference to all living things, whether they build that universe mentally or not, about how they perceive and sense it. All is perceived and detected using our five outer senses, and some other species have even more developed senses, which all receive stimuli through electrical impulses, meaning through electron clouds and clusters. When it comes to psychological worlds, it seems we have developed internal senses to feel and perceive them.

The universe can be perceived and sensed either way, whether the universe always existed out there, or if it is something we each build in our own mind. It is electrical signals reaching our brain that tells us what is out there, and as we will see from what follows, we are not all seeing and perceiving the same things.

6.7 Dreams and imagination are very similar to physical reality

We cannot deny that the physical reality is identical to dreams, and even to what we can imagine in our mind. For example, we can easily visualise the Palace of Westminster, the UK Parliament, in our mind, just like we could dream about it in detail. The only difference between these realities, is that the dream and our imagination are not relying on your five outer senses to detect or sense these realities. Both the imagined reality and the dream one, contain a physical reality identical to ours, where time and space might be a bit out of place and variable, but still very similar.

At the moment we wake up in the morning, we can't be sure if what we dreamt about was actually real, or just a dream, until our mind reassures us that it was not real. The distinction between imagined or dreamt realities, and the physical reality, is blurred. The chances then that the physical reality is just the same as mental ones are extremely high.

6.8 Mental health issues and hallucinogenic drugs are creating different physical realities

In order to verify the true nature of reality, to see if it is psychological and we each build our own bubble reality, like we build our dreams, we must review the most extreme cases of psychological disorders.

Whether they happen naturally, or after taking hallucinogenic drugs. Schizophrenia, hallucinations, visions, mass hysteria, those are the things we must explain.

Such events, as unsettling as they may be, support a reality which is a construction of our brain, much more than a real physical construction out there that always existed. When we have an hallucination, and I had two significant ones myself, after taking sleeping pills while really drunk, we see things that are not there. We build an entire reality in our mind that no one else shares, and yet, this is a reality that exists not only in our mind, but very much out there, just like the normal physical reality that our five outer senses detect and sense. It is quite an experience, because it feels so real, so real in fact, that it might actually be real.

In one hallucination I had, I was outside a boat peeing in the canal. My back was resting on the side of the boat, and I was holding a metal handle so as to avoid falling in. When my partner woke me up, I was actually in the boat, one foot on the bed, the other on the table, and I was peeing in the boat. On what was I resting my back then? Where was this metal handle I was clinging too, as if my life depended on it? How did I not fall on the floor and crack my head open, while I had one foot on the bed and one on the table? Was I actually truly outside in the reality of the hallucination, instead of inside the boat, until my partner woke me up, to bring me back to the collective reality that everyone else shares?

And when it is mass hysteria and thousands are witness to something impossible, or seeing something that is not real, they too share in another physical reality, as if interconnected to a wireless network, where they all saw the same impossible things. Like a multiplication of bread, fish and wine perhaps? And it was as real as our physical reality.

If our mind is capable of changing the physical reality on a whim, just like that, and to such a degree, and it all seems as real as the real thing, then there is not much difference between what our mind can create and the real thing. It supports entirely the fact that this physical reality, could be as psychological in nature as dreams and thoughts or imagination are.

6.9 Reality can be changed through will alone

A large number of people have tried to change their reality, and even the reality of others, by simply wishing for it, or praying, or chanting, or through hex and other things like black and white magic... and they succeeded. I have witnessed such changes personally, where buildings are different the next day, or trees or lamp posts are there, when they were not the day before. Or company logos and branding are different from before, followed by something radically changed in my personal life.

As mentioned earlier, I am writing a book about this called *Changing Your Future, Just wish it, be convinced and it happens.* Thousands of people have experienced the fact that reality is malleable through will alone, and dozens of books have been written on the subject. Don't you think if there was nothing there, it would easily be verified and forgotten?

By all means, give it a try. But careful what you wish for, as it might come true extremely fast, and not exactly as you expected. I was shipped to live in Los Angeles overnight, but it ended up being a complete disaster. It is safer to more generally invite into your life love, happiness, peace, health, wealth and freedom. The six things that you really need in order to be fulfilled, I would think, no matter the circumstances of your existence. Just repeat these six words, every once in a while, in your mind, without thinking too hard about it or anything, and you'll soon see the difference in your life. As easy as that, the simpler you keep it, the better it works.

The permanent and immanent physical universe, existing out there for billions of years, is incompatible with this ability to change reality, through concentrating and focusing on the changes we wish to operate. It is even incompatible with simple answered prayers.

Only a psychological reality, is compatible with this kind of malleability of our physical reality.

6.10 Ghosts and spirits are only seen and heard by certain people

You might not believe in ghosts or spirits, or consciousness that continue after death, in a reality of some sort. However, I heard somewhere that there would be enough scientific evidence to convince any jury in the Western world of the existence of such phenomena, if it were some long criminal case to be judged upon. The fact remains, that there is enough evidence to believe that there are ghosts, spirits and life after death.

Like for aliens and UFOs, the only way this can be denied, is through wilful effort to ignore the entire literature on the subject, the thousands of witnesses and their evidence, and the science that could support such facts. If our physics cannot explain any of this, then we still haven't found the proper physics that can explain our world.

Why would such phenomena be seen and heard by some, but not others? You have a ghost in front of you, half the people in the room will see it, the other half won't. Some will hear it, others won't. It is either there or not, and often only technology will detect it, see it, hear it, while humans won't, and vice versa.

This fact supports that we all have our own reality, in which we will only see and hear what our brain can actually receive, be it electrical impulses, codes, or other visual and auditory inputs of different kinds. And if such signals are out of our range, or the radio that is our brain cannot receive such frequencies, then we don't construct in our mind the complete reality others can see, hear and feel.

It tends to prove a psychological reality that we each build ourselves, where I will build that ghost in my reality, while others won't, because they have not received the signal. If a ghost was truly there in the permanent physical reality, everyone would see it, not only the technology that paranormal organisations have developed, to maximise such recordings of paranormal events. An interesting example here, is electronic voice phenomena, or EVP, that no human can hear, while voices are clearly recorded digitally. It tends to demonstrate that the ghost is there no matter what, in a permanent physical reality, that no one can hear or see, or that just some people will hear and see.

I believe it supports the psychological theory of the construction of the physical reality, as the ghost is also a living thing, with a mind, who also sends signals to the wireless network of the universe we are all connected to. And if our brain is attuned to the frequencies and intensities, of the electrical impulses of such signals, then we see it, otherwise we don't. Technology picks up on these signals, and we cannot deny here that it must all be electrical impulses, because our entire technology is based on receiving, emitting and recording electrical impulses of different frequencies and intensities.

6.11 High magnetic fields surrounding the brain distort physical reality

You might have heard of these experiments, like the one called the God Helmet, where a weak magnetic field applied to the brain can give you all the paranoia one might desire, including visual and auditory hallucinations of ghosts and spirits, and other religious or paranormal experiences.

Although this was used to disprove the reality of such spiritual and paranormal phenomena, as nothing more than distortions in the brain, it could be debated that, instead, it permits communication with the dead, and they are not then simply imagined. Either way, it does not matter whether this permits communication with the dead, or make you see and hear things that are imagined. It is hardly an argument that can be used to disprove the existence of ghosts, and it sorts of prove that reality appears more psychological in nature than not.

If a simple magnetic field applied to the brain can distort physical reality to such an extent, that you suddenly hear and see things that are not there, or that might be and that normally you would not see and hear, then to a large extent, our physical reality must be psychological in nature. How else could the brain so easily rebuild a different or distorted reality, that looks identical and feels the same as the real and permanent physical reality, if both were not psychological or mental constructs in the first place?

6.12 Autosuggestion and hypnosis can alter the physical reality we live in

Extraordinary feats have been achieved through autosuggestion and hypnosis. You can convince yourself or anyone of just about anything. You can build any reality you want by simply suggesting it, and it is believed, and it becomes the physical reality of the subject.

How could that be, if reality was not primarily psychological in nature? How could five senses convince you that you're a flying bird on Mars, without your brain stopping right there and thinking: wait a minute, this is not possible? How could you so easily rebuild an entire and real reality just like that? Granted the subject is still here, he is not flying to outside observers, he is not on Mars. But if you were to put under hypnosis the entire planet, and suggest to them that we are all on Mars, then what is the real physical reality we would really be living in?

Hypnosis and suggestions, are more than just imagining some new reality that seems as real as the real thing. As described in the book of Michael Talbot mentioned earlier, a subject can tell you the time from a watch hidden by someone, after they've been told that they won't see that person anymore for a while. Essentially, the subject no longer sees the person in his reality, hence he can now see the time on the watch behind the person he cannot see.

Reality is extremely malleable, and it is just our collective mind that keeps us all on the same page. We can see how easily we can derail, and start seeing pink elephants all over the place, and our mind certainly could not tell us what's real and what's not.

6.13 There is no distance in a psychophysical, virtual, simulated or holographic reality

We must admit, that possibly our physical universe is more a psychological construct of the mind, than a real physical reality out there, and could be formed by rapidly expanding from a point at the centre of our brain, every time we open our eyes or move to another room. And if somehow, our reality is more akin to a psychophysical reality than a real concrete one, then the physical reality becomes more like a virtual world, a simulation of the brain, or a holographic universe. And then, indeed, distances in the universe are meaningless, just like a thought, an idea or a dream, do not occupy any space in our physical world.

Anyone who believes we are living in a simulated world, a holographic universe, or a virtual reality, will need to consider the implications of such a world. The first one is, that there is no distance in this physical reality, just like we never really cover any distance in a computer simulation or a virtual reality.

This paves the way to imagine an entirely different way, by which to explain motion in such a world of expansion and contraction of matter, where there is no real distance to speak of. In chapter 7, we will now consider The Theory of Universal Relativity, called upon to replace Einstein's Theory of Special Relativity, as a theory of relative motion.

SEVEN

Universal Relativity Theory - The Atomic Expansion Theory of Relative Motion

FIFTH TYPE OF EXPANSION AND CONTRACTION OF MATTER, CONCERNING SPECIAL RELATIVITY AND RELATIVE MOTION

7.1 The Theory of Universal Relativity

The Theory of Universal Relativity is a theory of expansion and contraction of matter. Relative to the point of view, when objects or people move towards us or away from us, with motion, they simply grow or shrink in size. As if everything, including all the stars and the galaxies, were already here around us, and there were no distance.

Universal Relativity is a theory of relative motion, replacing Special Relativity. While Mark McCutcheon's Atomic Expansion Theory, replaces General Relativity and Quantum Mechanics, explaining the four main forces or interactions in physics: gravitational, electromagnetic, strong nuclear, and weak nuclear.

Both our theories complete each other, just like Einstein's theories of Special and General Relativity did. The difference is that our theories, although radically different from anything in science today, work at all scales, the very small and the very large. Together they are a Theory of Everything.

Whether objects moving away from us contract with relative motion, or are simply moving away in the distance without changing size, would not negate anything else I have stated so far in this book. It may seem like a crazy idea, but it is not so far off what Einstein himself was stating, in his Theories of Relativity. Objects truly do shrink in real time in Einstein, with acceleration close to the speed of light, and near very strong gravity fields, such as near black holes. But there are significant differences between Special Relativity and Universal Relativity.

So, why should I state that there is no distance in the universe, no speed, no velocity, that objects just relatively expand or reduce in size with motion? I have been asking myself that very question since the early 90s, when I first thought of it, and how no one would accept such a state of affair. It seems to be counterintuitive to what we see and experience in our everyday existence. However, since then, the idea that we might be living in a computer simulation or holographic universe, has become highly likely, and suddenly the idea is not so crazy anymore.

Nowadays, I feel that the more someone reads about relative motion, the more it seems that it is the only conclusion one can reach. And if what the New Age literature tells us is true, that instant travel anywhere in the universe is possible, then Universal Relativity must be correct. As it is the only theory out there that could explain it, being the only theory that claims that there is no distance or real space to begin with.

This is without resorting to wormholes of course, which would crush you anyway under any circumstance, despite what the Theoretical Physicist Kip Thorne is reported as having said, on his page on Wikipedia, in the section *Wormholes and time travel*.

Wormholes no longer exist in Atomic Expansion Theory, as there is no more warping of spacetime. If stargates truly exist, they don't open wormholes between gates, they must change the vibrational rate of matter, to match the interdimensional density of matter at the other end. Each location on Earth or in the universe, at any particular time, has its own unique frequency at which particles orbit and vibrate, which affects their size, and when we match that resonance, we instantly move there.

Is it really that counterintuitive? Is it not absolutely the same, if when your friend is walking away from you in the distance, he could instead simply be reducing in size from your point of view? Maybe this is just an equivalent analogy, and possibly this model of the universe, where people and things simply grow or shrink with motion, might lead to interesting maths and physics, but in the end, we all know this must be incorrect?

After all, is it not obvious that objects remain the same size, and simply move towards us or away from us in the distance? And that this expanding and contracting process, we witness with motion, is just a perception or an optical illusion, something children could easily believe, until they understand the world better?

Objects constantly expand, this inner expansion never stops and is independent of relative motion. A spaceship would go away, contract from our point of view, then come back and expand towards us. When it lands, we would witness that it continued to expand as normal while away, because it would be normal size upon its return.

We could say that it is only the effect of relative motion, that made it look reduced in size while away. Then, would it not be the same and simpler to say, that the ship only appeared to contract while away, simply because it was moving in the distance in space?

We don't have to think objects truly change their size with motion, since it is simply an optical illusion, due to the relativity of motion, and due to the limitation at which the speed of light travels. However, it is an equivalent way of looking at it, and in the end, it might be closer to the truth.

It makes me wonder why I would have thought something so outlandish in the first place, after decades of thinking about it, and visualising the world we live in. And why would such a vision of motion in the universe, be entirely supported in the New Age, spiritual and esoteric literature. Is everything not all expansion? Is there no real space, distance or time to speak of? Expansion is possibly the word that is used the most in the New Age literature, it is a key word that appears to explain everything. And yet this word is never truly properly defined.

What Jane Roberts has written about, is not that different from what most other channelled material is saying, or even, to some degree, from what the main religious books are telling us. If the physical universe is a creation, that it was thought into being by some higher power, then as physical as this reality might seem, it is still a psychological or mental construct. Psychological ideas don't occupy space, or certainly nothing like the space we are familiar with in our physical reality.

Even Mark McCutcheon, who, as far as I know, is not into New Age or religion, qualifies motion as ethereal in his book *The Final Theory:*

> "Motion alone is not a material concept, and can be quite ethereal when we consider that it is often completely arbitrary whether an object is in motion or not. As shown in the rethink of Newton's First Law of Motion in Chapter 3 [of the book *The Final Theory*], a given object can be in motion relative to one object but completely stationary relative to another; yet, an object in relative motion can make a very real impact when it strikes another object."
>
> Mark McCutcheon, *The Final Theory,* chapter 3

Could the physical universe truly be a mental construction? Perhaps developed collectively by all of us at the beginning of time, as Seth in Jane Roberts' books says, but individually re-created every day by each one of us, right from the centre of our brain, like computers connected to a network do? Could physical reality be constructed exactly like an idea we think in our mind, like the dreams we have every night?

Seth does state that ideas and dreams have physical form, they are made of the same stuff as our physical reality: electrons, atoms and molecules. And yet, Seth asks us: "does an idea or a dream take any space?". Are all the objects we imagine and dream about, cover huge distances? No. From that perspective, with motion, objects would simply expand and reduce in size from the point of view of each other, none of these objects would occupy any real physical space that would exist in our reality, even when made of real tangible matter.

I understand not everyone believes in some form of creation, religion, or even New Age or spiritual ideas. I cannot be sure myself, I am simply considering everything. However, on a more scientific level, there are also arguments to justify the non-existence of space and distance, or even speed with motion. Numerous books have been written about this world simply being a virtual reality, or a holographic universe, akin to what a computer would build in these elaborate simulations and games.

The difference with our physical reality, is that we are not limited to just electrons. Our reality is a world where electrons are bonding together to form clouds, clusters, protons, neutrons, atoms and molecules, as everything is made of electrons. And it won't take long for computers to be able to create something identical to our reality, as soon as we instruct electrons to form into these larger particles first, and programme our laws of physics into it.

This is assuming that this virtual reality would be a real physical one, observable and measurable with our outer senses. And not one that you would simply beam directly into someone's head, which could only be sensed from our internal senses. It would not matter, both realities would be identical in every way, made of the same matter, same particles, because in the end, it's a simple interpretation of the brain in both cases. In both of these worlds, there would be no space nor distance, and objects in motion would simply shrink or expand, compared to each other's point of view.

Our laws of physics must be the same everywhere, in all realities, dimensions or planes of existence. It is no longer just a matter of unifying the physics of the very small, of Quantum Mechanics, with the physics of the very large, of Special and General Relativity. Which Atomic Expansion Theory, and my new models for all these particles, have unified successfully at any rate.

It is also that our physical world might not be that different from the world of thoughts or imagination, or from the world of dreams, or from any type of virtual reality, or any other world of lower or higher density or dimension, or even the place where people appear to go when they die.

And here in this book, with Atomic Expansion Theory and Universal Relativity, finally we have laws of physics that can be applied to all these worlds, and that should work exactly the same way everywhere. A theory of everything entirely based on the expansion and contraction of matter, in a universe completely made of neutrinos composing electrons, protons, neutrons, atoms and molecules. Therefore, I feel justified in tossing aside our old concepts of size, motion, distance, space and time.

With relative motion, moving objects only ever expand and reduce in size. It is the same for light and radio signals. The source is simply shrunk from our point of view, and the clusters in light, and compressed bands of electrons in signals, simply expand towards us at a certain rate, which corresponds to the speed of light. And when they pass us and continue on their way, they simply shrink from our point of view.

Ultimately, it means that there is no space, no distance, and no speed in the universe. Everything we see is subject to the relativity of motion, therefore this kind of optical illusion quickly becomes our physical reality. And the New Age and spiritual literature have been telling us for a while now, that this reality is just an illusion, Seth calls it our camouflage reality.

We cannot be sure exactly of the nature of our physical reality. Thus, for the reasons mentioned above, and in the interest of developing a theory of everything that can be applied everywhere, it is best to consider that reality is more like a mental construction, a virtual reality or a holographic universe. Where everything is here at the same point, like it would be in a computer processor, before it is expanded unto a monitor, and objects only relatively expand or reduce in size with motion.

In this way, everything in physics can be explained by the simple expansion and contraction of matter. Everything is expansion, just like the channelled material has been telling us for a long time.

7.2 Fifth type of expansion and contraction of matter, and relative motion

There are little animations involving spaceships, with lights and mirrors, that you can check online. And fast moving trains, where it is impossible to tell which one is really moving, or which event is happening at what time, because they are only moving relative to each other. You must have come across these in school at some point.

Logically, it appears that motion must be relative to your frame of reference, because people don't see the same event taking the same amount of time to cross distances, depending on where they stand. This is called the relativity of simultaneity. From Wikipedia, it is defined as "the concept that distant simultaneity - whether two spatially separated events occur at the same time - is not absolute, but depends on the observer's reference frame".

As mentioned, the fifth type of expansion and contraction of matter concerns the motion of objects, which Mark McCutcheon has confirmed is relative, in his book *The Final Theory,* in chapter 3, *Rethinking Our Celestial Observations,* in the section called *Rethinking Newton's First Law of Motion.* He explains, in this important chapter, which describes the geometry of orbits according to Atomic Expansion Theory, how he views the relativity of motion:

> "A lone object cannot possess an absolute velocity or momentum - nor have any *specific* velocity or trajectory at all - without another object to serve as a reference point. Now consider the case where this object is speeding past a reference object (Figure 3-5 [20])."

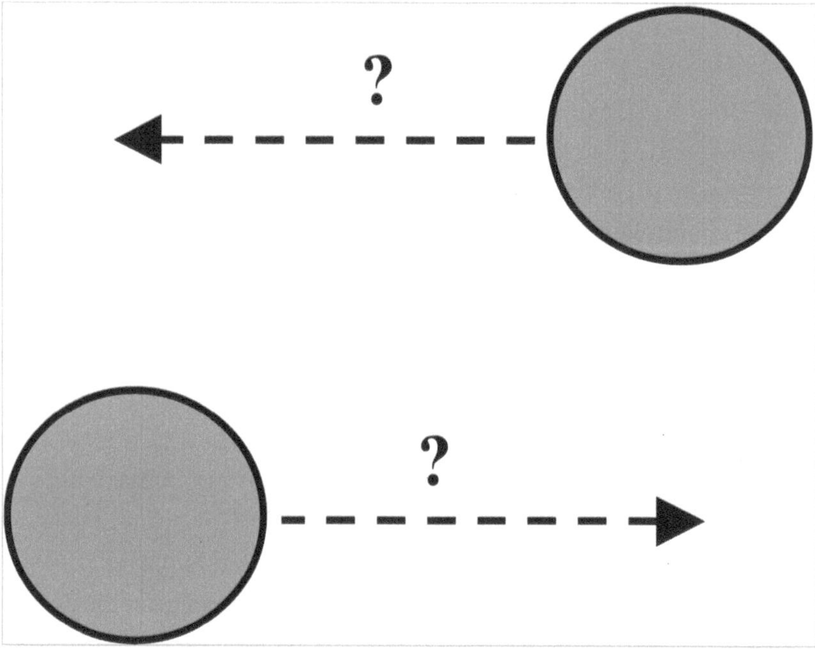

Figure 20 (3-5) One object speeding past another

"Which object in Figure 3-5 [20] is moving, and which is stationary? If these are the only two objects in the universe, and the square frame around Figure 3-5 [20] is not there for reference either, it is impossible to tell. If we consider one to be stationary then it is the other that is in motion, but this selection is completely arbitrary. In fact, it is not just that we cannot *tell* which is actually in motion, but that neither object actually *possesses* any *absolute* velocity or momentum at all; there is only *relative* motion between them. *Either* object could be considered to be moving while the other is stationary, or both could be moving toward each other at any of an infinity of speed combinations that add up to the overall speed of their approach toward each other. Both objects could even be moving in the *same* direction - say, to the right - with the one behind simply moving faster and thus catching up to the other. [...] Neither object individually *possesses* either an absolute velocity or an absolute state of rest."

Mark McCutcheon, *The Final Theory,* chapter 3

There are several theories of relative motion, Special Relativity being the main one in our physics today, and Universal Relativity the one I am proposing now. Mark McCutcheon's view, however, appears to be close to the Galilean relativity, from before Newton was incorporated into it, from whence it became known as Newton's relativity.

7.3 Galilean Relativity and Special Relativity

Here is the definition of Galilean relativity according to Wikipedia:

> "Galilean invariance or Galilean relativity states that the laws of motion are the same in all inertial frames. Galileo Galilei first described this principle in 1632 in his *Dialogue Concerning the Two Chief World Systems* using the example of a ship travelling at constant velocity, without rocking, on a smooth sea; any observer doing experiments below the deck would not be able to tell whether the ship was moving or stationary."

Galilean Relativity, Wikipedia

It happens that Galileo's relativity is also the first of the two postulates of Special Relativity:

7.3a First postulate (invariance of the laws of physics)

> "The laws by which the states of physical systems undergo change are not affected whether these changes of state be referred to the one or the other of two systems of coordinates in uniform translatory motion." "The laws of physics are the same in all inertial frames of reference." "That the laws of physics are invariant

> (i.e. identical) in all inertial systems (non-accelerating frames of reference)."
>
> "The theory is 'special' in that it only applies in the special case where the curvature of spacetime due to gravity is negligible. In order to include gravity, Einstein formulated general relativity in 1915. Special relativity, contrary to some outdated descriptions, is capable of handling accelerated frames of reference."

7.3b Second postulate (invariance of the speed of light c)

> "As measured in any inertial frame of reference light is always propagated in empty space with a definite velocity c that is independent of the state of motion of the emitting body." "The speed of light in free space has the same value c in all inertial frames of reference." "That the speed of light in a vacuum is the same for all observers, regardless of the motion of the light source."
>
> Special Relativity, Wikipedia

This means that Mark McCutcheon agrees with at least half of Special Relativity, since Galileo's relativity is its first postulate. I believe he has rejected all other postulates and other consequences of any theory of relative motion. This is what Mark McCutcheon wrote to me about both postulates of Special Relativity:

> "I also don't know that anyone has confirmed a constant speed of light across vast distances. It has been measured across extremely tiny distances on Earth, but I believe it's only assumption that this applies just the same across interplanetary distances and beyond?"
>
> "Well, I don't really see the need for the formality of 'postulate 1'. It's just common sense that the laws of physics are the same throughout the universe isn't it? Seems a bit of a silly 'postulate' to formalise like this.

> "Postulate 2 is just that... a postulation. Our limited lab testing on Earth can't possibly confirm such a universal claim given the high speed and short distances and limited test scenarios, and today's physics is so lost and confused that they can't reliably even think things through reliably in thought experiments either.
>
> "And no, I definitely don't agree with any of the claims of success, accuracy, verification of physical results of Special Relativity. I believe I addressed much of this in the book.
>
> "If you're after just a straightforward theory of relativity, it already existed well before Einstein. Galileo had the first formal theory of relativity, I believe. Einstein just screwed it up."
>
> Mark McCutcheon (in correspondence)

In the Atomic Expansion Theory of Relative Motion, objects start to considerably enlarge or reduce in size as soon as they move, and they don't need to move very fast. Just look at a friend walking away in the distance, shrinking away with motion, and this is what I mean. There is no need to reach velocities close to the speed of light to notice this effect, and as such I am not referring to Lorentz length contraction, that I explain below in the range of consequences of Special Relativity.

If you try to visualise this, moving away in the distance is equivalent to simply reducing in size, and then a moving one metre ruler would shrink as well, when you compare your size and your distance measurement, to the person moving away from you. And when that person looks back at you, from their point of view, it would be you reducing in size. And the question is, who is really moving away from whom, who is truly reducing in size? Because no experiment in physics you could devise could tell, even if you know that it is your friend that is moving away, and not yourself.

And if you search online concerning the relativity of motion, you will read about the Galilean and Lorentz transformation equations. Such equations are used for example to find out the right coordinates of objects, since the limitation of the speed of light distorts reality. Things in space are not exactly where we see them, or going at the speed we might calculate for them, because light takes time to travel to all observers, and its path along the way is affected by everything it encounters.

This is a very simplistic explanation of the relativity of motion, for a rather complicated subject. Lay persons and theoretical physicists alike, have been wreaking their brain around these concepts since the time of Galileo, and it is not getting any better. Especially since the weird phenomena of Quantum Mechanics have been introduced into our science.

7.4 Time is neither relative nor a dimension, as Einstein postulated in his concept of spacetime, and mass and energy are not relative

It is important to mention the greatest change from the Theory of Universal Relativity, compared with Albert Einstein's theories of relativity. Time is no longer relative, it ticks at the same rate everywhere, no matter the acceleration, and no matter any gravitational field, meaning how close a clock is to a massive astronomical body like a star. There is no more concept of spacetime interlinked together with mass-energy. This changes just about everything in physics.

In Special Relativity, the faster a spaceship goes, the more time runs slowly on the ship, compared to someone at rest. This is called *Relative velocity time dilation.* In General Relativity, the closer a ship is to a massive body like a sun, the more time on its clock will run slowly, compared for example to us on Earth. This is called *Gravitational time dilation.* Although, in both cases, on the ship they would notice nothing of the sort. Their time will never change its normal rate, so this is only relative to other observers. In my theory, relativistic effects are simple optical illusions, they don't have a real time effect on anything.

This will be greeted with scepticism, for anyone familiar with the actual state of our physics, because there are multiple proofs that time is in fact relative. The time rate of a clock on Earth is not the same as the one we will find on a GPS satellite. We calculated, that on a particular satellite, time runs 38 microseconds faster every day than a clock on Earth. And a clock on the space station, actually runs slower than a clock on Earth. This is a real physical and mechanical difference in the time rates, between three different frames of reference.

The time rate, on both the space station and on the GPS satellite, takes into account the speed they are going at around the Earth, which makes their clock run slower, as stated in Special Relativity. It also takes into account their altitude, or gravity, independently from their speed, which would also affect time, as stated in General Relativity.

A clock on the space station still runs slower than on the ground. While for the satellite in a higher orbit, and orbiting slower than the space station, time runs faster than on the ground. There are correcting software in all these clocks to re-adjust time, so it is the same as on Earth. It is considered that this proves neatly both theories of relativity.

There were also tests, where two identical very precise atomic clocks ticked at different rates, after one was flown high up in the sky on a plane, while the other stayed on Earth. And other experiments, where such clocks would run at different time rates, even when one is just 12 inches higher above the ground than the other, while both clocks are still on Earth near each other. Is this not proof enough that time is relative?

Not according to Mark McCutcheon, who states that time does not exist, and explains in his book *The Final Theory,* how we misinterpreted these proofs within our physics, saying that there is no need to invent bizarre concepts of *special or general relativity,* or even *time* itself as some mysterious entity or dimension.

He accepts the fact that, identical clocks could very well tick at different rates at different altitudes and with acceleration. He says that it is reasonable that this would occur if, for example, that device is undergoing ongoing 1g mechanical force on Earth, then absolutely no mechanical force in orbit. And with acceleration, more stress is put on these mechanical and electronic devices.

Here is a quote from Mark McCutcheon, from my correspondence with him, on the subject of these so-called proofs of the relativity of time:

> "Everything simply proceeds at its own unique mechanical rate, driven by the underlying expansion. A car does that. The countless systems in our bodies each do that. A simple mechanical alarm clock does that, all at slightly (or greatly) different rates, of course. And so do atomic clocks. And such highly sensitive, highly precise clocks are bound to show very sensitive variations depending on mechanical stress, such as 1g expanding planet on the ground against no such force in space, temperature variations (the manned space station has a very different temperature environment than a simple unmanned satellite), humidity variations, radiation variations, magnetic field variations, manufacturing differences, software differences, etc.
>
> "Then toss in the hugely artificial fiddling done with so called relativity correction software plus other miscellaneous 'corrections' as well (which are completely obscure and proprietary). The absolute last resort of complete desperation and confusion is to abandon such common-sense science and fly off on wild tangents of some mysterious 'time' dimension or phenomenon, and invent wildly flawed, fantastical 'general and special relativity' theories."
>
> Mark McCutcheon (in correspondence)

There are other arguments against time being relative. The first one, almost all sources of channelled material in the New Age and spiritual literature, cannot stress enough the very same statement of Mark McCutcheon, that time does not exist. How can something that does not exist, which is something we have invented as a practical convention in physics, for our measurements and equivalence calculations, suddenly goes on to become a dimension, to be linked into a spacetime continuum, as Einstein's theories depend on? The New Age is light years away from our actual science.

And finally, the ultimate argument for me, which made me reach the very same conclusion, while writing my Theory of Universal Relativity, is that my own research seems to confirm it. All electrons and atoms are constantly expanding at a constant rate, independently from all frames of reference, or from relative motion. Nevertheless, with relative motion, the limits of the speed of light distorts what we see, bringing all these relativistic effects on size, distance, location, speed and direction of motion. We cannot be sure where exactly in the universe something is really located, and its true speed or velocity. And this is why we need highly accurate transformation equations to give us some indication.

These variables just mentioned, are the visual ones that could be distorted by the limit of the speed of light in its travel. They are simply an optical illusion, brought on by relative motion. What we see is not what is actually happening in the universe, unlike what Einstein thought. Relativity only gives the illusion, that these values are changing from one frame of reference to another.

Time, mass and energy, are not visual variables in our physics, their value cannot be affected by the speed of light taking time to reach us, and be different for observers in different locations. Hence time, mass and energy are not relative, they are constant and remain the same in all frames of reference.

In Special Relativity, if a spaceship travels everywhere in the universe at all sorts of speeds, and comes back to Earth some time later, only the value of time would have changed. Time is the only variable which undergoes a real time relativistic effect in Einstein's theory. Even relativistic mass and energy has no real relativistic effect, because the ship will have the same mass and energy before and after its trip in space, minus what it spent in its travel. There is no reason why time should be any different.

This is where it becomes a bit confusing. Because mass and energy, in this case the kinetic energy of moving objects or mass, including light, are subject to relative motion. And the momentum or force of impact of any mass, meaning its energy, does change with the geometry of expansion, like through the slingshot effect, the natural orbit effect and gravity. And this is due to relative motion.

However, mass and energy are not relative values that will change for observers in different locations, which is what is meant by light, and anything affected by light, being relative to the frame of reference. A mass will acquire more speed, and hence more kinetic energy, due to the geometry of expansion, in the same amount for all observers, no matter their vantage point. The force of impact of a certain mass, when it hits something, even for the momentum of light, will be measured to be the same for everyone everywhere, no matter what they see and when they see it.

Remember the twin paradox. The twin who would have remained on Earth, would have aged faster than the one on a spaceship, since on the ship time would have ticked slower. How could this be possible, if relativity is all but an optical effect? It might seem like common sense, that your twin might age slower than you, just because he accelerated and went close to the Sun in his 20 years in space. But not from the point of view of even Einstein's relativity of motion. Because as stated earlier, it is not just that we cannot *tell* which twin is actually in motion, but that neither twin actually *possesses* any *absolute* velocity or momentum at all, as there is only *relative* motion between them. You can safely pack and ship your twin into space for 20 years. When the ship comes back, your twin won't make you look ancient, you will both have aged the same.

Mark McCutcheon has debunked this twin paradox, along with the debunk of Standard Physics. The direct links are at the end of chapter 1, at section 1.9.

Thus, we must conceptualise a theory of relative motion, where time does not vary from one frame of reference to another. Albeit, it could still vary very slightly, due to physical mechanical stress and other reasons, as mentioned in Mark McCutcheon's quotes above.

In essence, time does not really exist in the universe or in our physics, except as a useful convention. This said, we cannot escape the reality that time is actually a very useful and necessary concept, in a world where every fraction of a second must be accounted for.

After saying all this, I must point out that time is still relative, although not the way Einstein envisaged it. I will cover in chapter 9, the concept of interdimensional time and higher dimensions, where time is still relative to the density and vibrational rate of matter.

Essentially, the faster all our particles are orbiting within us, or within an environment, the faster we live and that events take place. It's like living in fast forward instead of slow motion, but only from the point of view of others not running on the same time rate. Otherwise, everything seems normal, within the dimension or plane of existence we find ourselves in.

7.5 $E=mc^2$, and the range of consequences of Special Relativity

This is the range of consequences of Special Relativity according to Wikipedia:

> "Special relativity implies a wide range of consequences, which have been experimentally verified, including length contraction, time dilation, relativistic mass, mass-energy equivalence, a universal speed limit and relativity of simultaneity. It has replaced the conventional notion of an absolute universal time with the notion of a time that is dependent on reference frame and spatial position. Rather than an invariant time interval between two events, there is an invariant spacetime interval. Combined with other laws of physics, the two postulates of special relativity predict the equivalence of mass and energy, as expressed in the

mass-energy equivalence formula $E=mc^2$, where c is the speed of light in a vacuum."

Special Relativity, Wikipedia

As mentioned previously, time has no special status in Atomic Expansion Theory, nor is it a dimension all of its own, along with length, width and height, so they can together form a so-called spacetime, and affect each other with motion. There is no more concept of spacetime, time does not really exist, except as a convention, so there is no more universal light speed limit imposed upon us. Nothing stops us from going faster than the speed of light.

Moreover, the speed of light itself varies with gravity, partial orbits, and everything it encounters in space. Everyone can see light coming towards them no matter their speed, even multiple times our actual measurement of the speed of light. Also, observed events might move faster or in slow motion, depending on one's speed and direction of motion, towards or away from the light source. I will discuss visualisations of all this, later in this chapter.

The equation $E=mc^2$ has been reviewed in light of Mark McCutcheon's Atomic Expansion Theory. Energy is mass, it is made of electrons. Consequently, mass in that equation truly means atomic mass, or the mass of the numerous electrons inside the atom, including the ones composing the protons and neutrons.

As soon as these electrons are outside the atoms, they usually form into clusters, for light and radiant forms of energy, and clouds for magnetic fields, at which point they are considered energy. Energy means electrons freed from the atoms, still all interconnected within the subatomic realm, or expanding freely into space.

This equation compares the same thing, electrons in two different forms. As a consequence, of course there is an equivalence between them. The only difference really, is the speed at which that matter goes. And it would not matter, whether that matter was in the form of clusters of electrons, or fully formed within atoms and molecules. The impact when going at the speed of light, would be the same. If you accelerate an asteroid to the speed of light, once it hits something, the impact will be equivalent to a nuclear bomb. And there is an excellent reason for this, Einstein's equation is simply the classic kinetic energy equation for moving objects.

For Mark McCutcheon, the speed of light c refers to the higher expansion rate of the electrons, after the crossover effect, which defines their speed once they are expanding freely into space. Nevertheless, he stated that we cannot be sure at this time if the speed of light is constant.

In my opinion however, c is not equal to the speed of light as we have recognised it, 299,792 kilometres per second, or 186,282 miles per second. Nor is the expansion rate of the electron clusters different from than the one of the atoms. So, the expansion rate cannot justify the high speed of the electrons, once they are outside the atoms freely expanding into space.

The speed of light varies, there are other variables to consider when calculating the speed of light or c. They are the speed at which the clusters of electrons are launched out freely into space, which depends on the excitation and speed of the orbiting neutrinos and electrons, and which accounts for most of the speed of light. The speed of light also depends on the size of the clusters of the electrons at the source, their expansion rate, which is the same as the one of the atoms, the distance to be covered until destination, and the size and expansion rate of the object at the destination.

All of the variables, except the initial speed at which the electron clusters are launched out of the subatomic realm, are the ones from the Atomic and Subatomic Expansion Equation to calculate gravity. Because the speed of light, or c, is to be calculated like gravity, a changing distance between two expanding objects, one of these objects being the electron clusters in light, or the size of the overall beam of that light. The square ^{2} in the equation, means that the speed at which the electrons travel, accelerates. Since in gravity, the distance reduces as an acceleration.

To summarise, that initial speed of the electrons, as they are freed from the subatomic world, is not the speed of light, since to this initial speed, we must then calculate the speed of light as gravity, using the Atomic and Subatomic Expansion Equation. The acceleration and variation of the speed of light, have already been discussed in detail in the first three sections of chapter 5.

And if you still wonder whether or not light can be affected by gravity, please note that this entire section is precisely to show that the clusters of electrons within light, are no different from any other objects going at the same speed. The force of impact, or the momentum for any atomic object or subatomic object, including the clusters of electrons in light, use the very same equation, as we will shortly see.

In chapter 6 of his book *The Final Theory,* the section called *E=mc^2: What is Energy-Matter Conversion?,* Mark McCutcheon shows that "Einstein's famous E=mc^2 equation does not refer to a mysterious 'energy-matter conversion' process, but simply the classical kinetic energy of free electron clusters". The classical equation for the kinetic energy of a moving object, that existed prior to Einstein, is $E_k = \frac{1}{2}mv^2$, which states that the object's kinetic energy, E_k, is equal to one-half of its mass, times its velocity squared. Here is what Mark McCutcheon says:

> "A number of derivations can be found for Einstein's E = mc^2 equation; and, while the more esoteric among them can become quite involved, the fact is that this equation can be easily derived in a few lines of simple classical equations of motion, as shown in the following four-line derivation:
>
> 1. p = E/c - Momentum of light, p, equals its energy content, E, divided by its speed, c
> 2. p = mc - Momentum of light, now stated in terms of its mass, m, times its speed
> 3. E/c = mc - Equating lines 1 and 2

4. $E = mc^2$ - Rearranging line 3 gives Einstein's well-known equation

"The first two lines above are simply two common equations for the momentum of light from classical physics (physicists traditionally denote momentum by the letter p). Line 1 shows the momentum that light imparts to materials that it strikes, stated in terms of its energy content divided by its speed. This is confirmed by experiments in which light is shone onto material from which it does not reflect back. It is worth noting, for reasons that will become clear shortly, that the momentum expression in Line 1 increases for materials that are more reflective, eventually doubling to the expression $p = 2E/c$ when the incident light is fully reflected from a mirrored surface.

"Line 2 shows the momentum that light would have according to the classical momentum equation for all moving objects - namely, momentum equals mass times velocity, or $p = mv$, with the velocity, of course, being the speed of light, c, in this case. It is interesting to note that, today, Line 1 is considered the proper physical description of light's momentum, while Line 2 is considered the more abstract expression since light is believed to be speeding energy - not mass - in today's science. In contrast, Expansion Theory shows that Line 1 is actually the abstraction and Line 2 is the reality, since light is actually a beam of physical matter particles (electron clusters) and not 'pure energy.' Regardless, since these two lines describe the same property, they can be equated to each other (Line 3) and rearranged (Line 4) to give Einstein's famous equation, $E = mc^2$."

Mark McCutcheon, *The Final Theory,* chapter 6

Mark McCutcheon demonstrates that $E=mc^2$ is simply derived from the classical equations for the momentum of moving objects or light. Light is actually a beam of physical matter particles (electron clusters) and not "pure energy". Light has its own ongoing expansion force from within, continually and forcefully pushing it onward as always (momentum).

Hence there is nothing special about this equation, it existed well before Einstein. Mark McCutcheon even modified it to be more accurate, once he understood where it came from and what it means. The real value of the kinetic energy for light, depends on if what it hits is a reflective surface or not, as clusters of electrons bouncing off a mirror can double their force of impact. And depending on how reflective all the surfaces are, which will be hit by the clusters of electrons, the correct Einstein's equation to measure the kinetic energy of light, should be anything between $E_k=mc^2$ for no reflection, and $E_k=2mc^2$ for full reflection.

There is one thing that could be puzzling, if the speed of light is actually accelerating with gravity. The kinetic energy at the destination, or the impact of the momentum of light, should be greater the longer it takes for the light to arrive, and the longer the distance covered. This is because the speed of light constantly accelerates with time and distance. It seems counterintuitive, light does not seem to have a greater impact the longer it takes to get somewhere, it's quite the contrary in fact. If you are near a source of heat or of light, you will feel it much more, the closer you are to that source.

Quite possibly this is because in time, most of the clusters of electrons in light dissipate, as they expand and travel across all sorts of particles they meet along the way. Such as gases, magnetic fields (clouds of electrons), neutrinos, molecules of air, etc. Light might very well stop accelerating, or vary in its acceleration, due to terminal velocity, as it meets resistance during its travel through all sorts of material. If a light source is weak, it won't go very far before it dissipates, which is why people further away won't see that light. It is like heat. Heat, like light, is just clusters of electrons, although larger ones. If you put a heater outside in winter, the heat would not go very far before it completely dissipates.

As a result, the momentum of light might very well hit an object harder, the longer it takes for that light to arrive at destination, but only on short distances. This is provided, that light does not also stop accelerating at some point because of reaching some sort of terminal velocity, where any frictional forces or resistance encountered, impedes on its acceleration. As soon as a light source sends clusters of electrons freely into space, or a radio signal, eventually the clusters in light, or the compressed bands of electrons in the radio signal, disperse and lose coherence, and the light or the signal become weak with time and distance.

As you can see, measuring the true speed of light is no picnic. And neither is figuring out exactly where objects truly are in space, when light beams travel in a twisted way, deviating as they go into partial orbits around larger objects, such as the Earth, the Sun and even galaxies. Especially when light would vary due to gravity and the slingshot effect, due to the geometry of expansion, even in a vacuum.

7.6 Length/Lorentz contraction, and the optical illusion due to the time it takes for light to travel

Length contraction according to Wikipedia:

> "Length contraction is the phenomenon of a decrease in length of an object as measured by an observer which is traveling at any non-zero velocity relative to the object. This contraction (more formally called Lorentz contraction or Lorentz–FitzGerald contraction after Hendrik Lorentz and George FitzGerald) is usually only noticeable at a substantial fraction of the speed of light. Length contraction is only in the direction parallel to the direction in which the observed body is travelling. This effect is negligible at everyday speeds, and can be ignored for all regular purposes. Only at greater speeds does it become relevant. At a speed of 13,400,000 m/s (30 million mph, 0.0447c) contracted

length is 99.9% of the length at rest; at a speed of 42,300,000 m/s (95 million mph, 0.141c), the length is still 99%. As the magnitude of the velocity approaches the speed of light, the effect becomes dominant [...]."

"In this equation it is assumed that the object is parallel with its line of movement. For the observer in relative movement, the length of the object is measured by subtracting the simultaneously measured distances of both ends of the object. For more general conversions, see the Lorentz transformations. An observer at rest viewing an object travelling very close to the speed of light would observe the length of the object in the direction of motion as very near zero."

Length/Lorentz contraction, Wikipedia

When objects appear to contract with motion, as in the length contraction or Lorentz contraction, it is only an optical illusion due to the fact that what we see is limited by the speed at which light travels. What we see is not to be taken literally as Einstein states, a spaceship does not really contract with speed. The relativity of motion plays tricks on us.

As a proof of this, when a spaceship travels in space at nearly the speed of light, and that from our point of view its length has contracted, and its time rate appears to have slowed down or nearly stopped, and its mass is reaching critical proportions, that would require an infinite amount of energy according to $E=mc^2$, in order to reach the speed of light; to the passengers on the spaceship everything is just fine. They don't see their length being contracted, nor their time running any differently than on Earth. And certainly, they would not feel so heavy or massive that they would require so much more energy to suddenly go just a little bit faster, and cross over the speed of light threshold. From their point of view, the measurement of their speed of light, would be different than from our measurement as seen from our own point of view.

All these extreme values for our variables in physics, that we observe from our frame of reference, when objects elsewhere move at a speed closer to the speed of light, are due to the relativity of motion, which also affects light, and it is all just an optical illusion. In reality, in its own frame of reference, that spaceship and its passengers are just fine. With a little bit more energy or thrust, they could go faster than the speed of light, from our point of view. Their maximum speed, is not limited by what we see from here due to relative motion, and their mass has not reached a disproportionate level either.

And neither particles, in particle accelerators, become so massive as they near the speed of light, that it would require an infinite amount of energy to push them further. The problem is, what we are using to move them around, electricity and magnetic fields, is made of electrons that can only travel at the speed of light, hence the difficulty in getting these particles to go any faster. There is no spacetime, where time is so linked to space, that what we see is actually what is real; nor is there a mass-energy relativistic effect. I've read somewhere, that many physicists in their experiments, even in Standard Physics, have already come to that realisation.

Since with the relativity of motion, we cannot tell if it is the spaceship, or us on Earth, which is moving near the speed of light, then what would the people on this ship actually see, when they look back at us, even though we are not moving at such a speed? They should see exactly what we see, our length would have contracted from their point of view. Hence, it is obvious that this is just an optical illusion, and nothing truly contracts in reality.

It is only when we compare their frame of reference to ours, that relative motion makes them appear as we see them. This is the reason why the configuration of the universe as we see it from here, is relative to our point of view, and could look different once we venture into space at high speed, especially at speeds faster than light. This is all an optical illusion, due to the limitation of the speed at which light travels, which takes time to arrive at all destinations.

The time rate of the people on the ship has certainly not stopped, as if they were stuck in a black hole. There are no black holes in Atomic Expansion Theory, they are merely massive dead stars that went supernova some time ago. And dead stars simply don't emit light, even if they can still orbit and influence other astronomical bodies, following the natural orbit effect of the geometry of expansion.

Most people have heard that time is relative, that it changes from one frame of reference to another. But people are less aware that distance is also relative. Not only is there a time dilation process, but there is equally a distance dilation process in Special Relativity, and also in Henri Poincaré's relativity, from before Einstein.

Length contraction is one of the consequences of the postulates of Special Relativity, as stated above. It is the source of my Theory of Universal Relativity, it is why I thought of it in the first place, when I was a teenager, years before I read the book *The Final Theory* of Mark McCutcheon. Although this length contraction, as specified above, is solely an optical illusion, due to the time it takes for the light of an event to travel to us. It is not really happening, as Einstein thought. Still, it made me realise that if objects were simply expanding or shrinking, from the point of view of different observers, instead of moving and covering distances, suddenly a lot could be explained, without the need for Einstein and Minkowski's spacetime concepts. And then, everything in physics could be explained through the simple expansion and contraction of matter.

Also, Mark McCutcheon has rejected that we would even see what Lorentz states here, for objects going at near the speed of light, where they would actually nearly disappear from view. He may have a point, I am not sure either if we would see any extreme length contraction, for objects going at near the speed of light. It is not like we have seen any object going that fast in the night sky, so we could verify it. There is nothing that we have observed so far in the universe, at the larger scale, which is made of atoms and molecules, moving anywhere near the speed of light. Except, perhaps, for UFOs.

And nor is this particularly surprising, according to Mark McCutcheon, considering the mechanics of the universe.

It does not mean that we cannot go faster than the speed of light, there is nothing now in our physics preventing us from doing so. However, by the time we figure out the details of what I'm saying in this present book, we won't need to attempt to break the light speed barrier anymore, in order to travel instantly anywhere in the universe. It will be like Neo in *The Matrix,* who once he's ready, he'll no longer need to dodge bullets.

Travel is now a question of finding ways of changing the vibrational rate of our matter, which also defines our size, and matching it with the ones of the locations and timeframes we wish to visit. It may be in the past or the future, 10 miles away, another planet or galaxy, or realities of different densities, in lower or higher dimensions, or in other planes of existence.

As the Bible says, the dead will rise. As soon as we develop the proper technology that is, because they were never dead to begin with. To my utter dismay, it seems that we live forever. But as Freddy Mercury sings, who wants to live forever? It might be a good thing that time has no meaning out there, or that everything happens at once, in a fluctuating timeline. More about that in chapter 9.

7.7 It is not time that is relative, it is the speed of light, as only matter can be affected by relative motion

There are things Einstein said in his theory of relative motion that were correct, but for the wrong reasons. In this section, I hope to demonstrate that it is not time that is relative, it is the speed of light, and as a consequence, the speed of light cannot be constant for all observers in different locations, even if somehow in average the same speed for light is recorded in these different locations. We may or may not all measure the speed of light to be close to c, 299,792 kilometres per second, or 186,282 miles per second, but it is not due to time being relative.

Light is made of clusters of electrons moving freely into space, granted, with a speed limit, the limit at which light travels. But only matter, objects or particles, can be affected by relative motion. Time does not move, it is not a material physical concept of any kind, how could it be affected by the relativity of motion? The only way to explain what is going on, from different viewpoints, when it comes to light and what we see, is if it is the speed of light that is relative.

So, what's the difference? Well, it means that you may very well measure the speed of light to be near c, while sitting here on Earth, but I would also measure the speed of light to be near c, while being on a spaceship going at twice the speed of light. This is due to relative motion alone. And it means that, my speed of light is actually three times yours, even if relatively speaking, I measure it to be the same as yours. How is this possible? This is what I will attempt to explain in this section.

Einstein stated in his second postulate of Special Relativity, that the speed of light in a vacuum is the same for all observers, regardless of the motion of the light source. While Mark McCutcheon believes, that we don't know if the speed of light is constant in all frames of reference, even in a vacuum, and he does not believe so. But it is only in a vacuum that the speed of light should be constant, according to Einstein.

To summarise the entire discussion so far, at the risk of repeating myself for the purpose of this section, in practical terms, space is never empty nor a vacuum, just about everything affects the speed of light. After its initial launching speed, light going through water, or molecules of gas in space or on Earth, passing through a magnetic field and other clouds of neutrinos, all of this slows down the speed of light.

Even in a vacuum, the speed of light can be affected, as it is being deflected and accelerated while passing by near expanding moons, planets, stars, solar systems and galaxies. This is due to the slingshot effect of anything entering into an orbit, or a partial orbit, around something much larger than itself. This accelerates the speed of light, and deviates its path, on top of its acceleration due to gravity. This said, gravity and orbits are related, they are the same phenomenon, in the sense that they are both the result of the geometry of expansion. Overall, I doubt very much that different observers would all measure the speed of light to be near c, even in average, but who knows when it comes to relative motion?

Today we measure the speed of light by how much time it takes for a beam of light to cover a certain distance, and that's it. For example, light will take around eight minutes to cover the distance between the Sun and the Earth. But there is a better way to visualise and measure this.

When we measure the speed of light, what we are actually measuring is the changing distance in a certain amount of time, between two expanding objects. Usually between light, which is made of clusters of electrons, and another object usually made of atoms and molecules, like the Earth. It would be an equivalent way to measure that speed of light, and one which could be more accurate, if we could devise the right equation, and if we knew all the variables involved.

We have such an equation already. The changing distance between expanding objects is gravity, calculated using the Atomic and Subatomic Expansion Equation described in the gravity section in chapter 3. The D' in the equation, is the changing distance per second between two expanding objects, or of a bunch of particles, of a certain radius, hence of a certain size. And now, we even know the expansion rate of electrons and electron clusters, it is the same as the Universal Atomic Expansion Rate as defined by Mark McCutcheon.

I see no reason why the clusters of electrons within light, or bands of electrons within radio signals, should be a special case in our physics, which would render them immune to gravity. It is already clear from the Gravitational Lensing effect, that the path of the light in space, deviates when passing near large expanding objects, such as a sun or a planet, but that light is moving too fast to enter into an orbit around them. It is certainly being caught in a partial orbit which deviates its course, and consequently, most likely affects its speed.

However, we must also add the initial speed at the source, at which these electron clusters and bands of electrons are launched freely into space, which is quite a high speed. Most of the speed of light, is a result of this initial speed, required for these electrons to free themselves from the atomic and subatomic realms.

And since gravity is an acceleration, the speed of light and everything in the electromagnetic spectrum, must be accelerating. I don't know how anyone on this planet could accept this statement in this day and age. I can only imagine all the proofs out there that are against such a statement, and yet I am absolutely convinced that it must be true. I hope someone will soon prove it.

For a same source of light, the speed of light will be different for multiple observers located in different locations, as the speed of light varies and is relative. Let's visualise this in two thought experiments, in the next two sections, one with a moving source of light while we are at rest, the other with a fixed source of light while we are in motion.

7.8 Visualisation of a moving source of light on a spaceship, travelling faster than the speed of light, as seen from Earth and another ship

Let's imagine a spaceship passing through our solar system at twice the speed of light, compared to our own speed in space. Then, they decide to light up a huge beacon of light, illuminating their entire ship, that we could see from Earth.

On the ship, their calculation for the speed of light from that same source, could vary a great deal from ours. They are already going at twice the speed of light, the same speed as the source of that light. Consequently, they only need to consider the speed at which the electron clusters are launched freely into space, and gravity, the decreasing distance between them and the beacon.

But it is the same for us, this is all we would need to do. Even if these electron clusters within light were already going at twice the speed of light, to which we must add the added high speed at which they were launched out of the atomic and subatomic realms, from there on it is just a calculation of gravity, a changing distance between Earth, and moving and expanding particles.

To be honest, I'm not quite sure of what we would see from this light source. My guess is that light would travel faster, if the ship was coming towards us at twice the speed of light, and we would see the ship moving in fast forward. If the ship was moving away from us, at twice the speed of light, light would arrive slower, and we would see that ship in slow motion. I think this would be true whether or not the speed of light is nearly constant everywhere.

It is possible that it is not the speed of light that would change, and that the speed of light would be c no matter what. I don't think this is likely, since these particles were already going at twice the speed of light when they were launched into space. They have not suddenly stopped in midstream to adjust to our own speed in space. Instead, it is possible that the speed of light might appear constant for people in different locations, due to relative motion. It does not mean that the speed of light is the same everywhere, as it is actually radically different everywhere. It only means that we relatively measure it to be the same, because of relative motion.

We would also need to consider any magnetic field or matter such as gases between us and the beacon, and other planets, moons and sun that could deflect light, anything that could slow light down or speed it up. In this scenario however, we will ignore anything that could affect the speed of light or make it deviate from its path.

We could use Newton's Law of Universal Gravitation, instead of the Atomic and Subatomic Expansion Equation. Maybe we could devise a general mass of the electron clusters in light at the source, then again, what would be the other mass at the other end? Earth's mass? Possibly that would work.

If Earth were suddenly ten times its actual size, it would expand considerably and proportionally, in the time that it would take for light, to go from the beacon on the spaceship all the way to Earth. It would affect the speed of light, it would arrive earlier than on the smaller Earth.

However, let's not forget that the speed of light is an acceleration, and it would continue to accelerate for the further distance it would have to cover to reach the smaller Earth. In the end, possibly the speed of light would be the same no matter the size of the planet, in this particular thought experiment. And this acceleration of the speed of light could lead us to believe that the speed of light is constant, when in fact it is accelerating.

If Earth was suddenly ten times its mass and size, even in Newton, we would obtain a result where the speed of light would arrive earlier than on the less massive Earth. Nevertheless, the calculation could not be as accurate as with Atomic Expansion Theory. A planet made of Styrofoam instead of rock, or made of a radically lesser density, would have a very small mass compared with the one made of rock. Newton's equation would bring radically different results for the speed of light, based on mass measured from centre to centre of the objects, while both planets would be the same size and in the same location.

Which is why the Atomic and Subatomic Expansion Equation, based on the size of the objects, their expansion rate, and calculated from surface to surface, brings more accurate results. Gravity cannot depend on the mass of objects, as in Newton, assuming we could actually guess correctly the mass of any object, without knowing its density. And even then, we would need to know if the density is the same across the entire object, which on the Moon, for example, it is not. The Moon is denser near the visible side, and less dense on the hidden side. Its centre of mass, from where we must calculate the direction of its expansion, which will cause gravity to vary all across the Moon, is nearer us. We don't need to worry about any of that, when we only consider the size and the expansion rate of the objects, which is not affected by mass or density, when not measured on the ground.

Calculating gravity or the speed of light using the Atomic and Subatomic Expansion Equation, would show that, for the people on the spaceship standing near their beacon, the size of the electron clusters on such a short distance, between them and the beacon, would mean that they would calculate the speed of light as normal, around c. And this is despite the fact that they are going at twice the speed of light compared to us. This is what I mean by the speed of light is relative to the frame of reference, or viewpoint.

The velocity of any observer does not matter, the changing distance or the speed of light will depend on the expansion rate of the electrons in light from the source, and the radii of the two objects between which we are calculating the changing distance. Hence, a spaceship going faster than the speed of light, would not suddenly be unable to see within their own ship, light would not freeze in mid-air.

This is important, because if we ever go faster than light, there will be no issue with light. Light can travel faster than the speed of light, relatively speaking, relative to other observers in other locations. Because light will always simply depend, on the changing distance between expanding objects or observers, in different locations. And this speed varies greatly from one frame of reference to the next, as well as being an acceleration.

In order to make it easier to see what would happen to electron clusters in light, from the beacon, once they are freely launched into space at their initial speed, let's assume the clusters are instead clusters of asteroids all sensibly of the same size. Normal gravity between atomic objects, such as asteroids and Earth, would mean that they would continue to expand, and the distance to decrease in an acceleration, until the clusters of asteroids hit Earth. It would be exactly the same thing for the speed of light, for those clusters of electrons.

The frequency of that light would also make a difference in its speed, since the clusters would be either smaller or larger depending on the frequency. But, do we measure the radius of each cluster, or asteroid, or of the overall collection of electron clusters or asteroids? And would overall beams of light, differ in size depending on the light frequency? This is important in defining the gravity or its speed.

The frequency of light defines colours. Different light colours or frequencies, cannot travel at the same speed, since their clusters are not the same size, and the Atomic and Subatomic Expansion Equation takes size into consideration, although the difference might not be that significant from our point of view. But the size of the overall beam of clusters, no matter the size of the clusters, might still be the same, no matter the frequency of the light.

However, as it takes more energy to release smaller clusters of electrons from the subatomic realm (hopefully this is a correct assumption), for example smaller clusters in ultraviolet light compared to larger clusters in infrared light, it is likely that the smaller clusters are ejected faster than the larger ones. Hence, the initial speed at which the clusters are ejected into space, might differ with the frequency of the light. As can be seen from this discussion, measuring the speed of light is not as easy as measuring how long it takes for the clusters to go from A to B.

If I am correct, in this way it could be said that the speed of light is relative. It is possible that both on the ship and here on Earth, we would measure the speed of light to be c, from that beacon going at twice the speed of light. But in reality, relatively speaking, on the ship they are going at twice the speed of light, and their speed of light could not be the same as ours, even if they would measure it to be around c as well. This is due to relative motion.

We can see that the electron clusters would start to expand from the moment they are freed from the subatomic world. No matter the speed of the light source, the only way to calculate the speed of light between different observers and the light source, is through calculating the changing distance every second between two expanding objects, which is gravity, whether they are made of atoms or electrons. But I would say that these clusters are moving at three times the speed of light at launch, therefore the speed and direction of motion of the light source could matter a great deal.

If a second spaceship is moving further away or towards the beacon, then every second, the distance changes between these two ships. And if they are expanding while space between them is not, it would change everything even further, affecting the calculations of their respective measurement for the speed of that light, and their measurement relative to each other.

Depending on the point of view, I suspect not everyone would measure the speed of light to be around the value of c, depending on their velocity and direction of motion, especially when going faster than the speed of light. With this option, the speed of light of each source in the universe, would be travelling at radically different speeds. Although, it is equally possible that the speed of light would be around c for everyone, and for every light source, no matter their speed and direction of motion. But the images they would see, might be going in fast forward or slow motion anyway, or even freeze if they are moving away from a source at the exact speed of light, all due to relative motion.

Testing would be required to verify and confirm all of this, and even then, I'm not sure we would be able to tell which option it is, and what is really happening here with these clusters of electrons. But comparing them to fired bullets from a gun and moving asteroids, should help us in our visualisations and thought experiments.

7.9 Visualisation of a fixed source of light, the Sun exploding, seen while travelling back and forth, at slower and faster than the speed of light

Much has been said of Einstein's thought experiment, of what it would be like to run at the speed of light alongside a light beam, for example showing the explosion of the Sun. For a second, he thought the image would freeze in time, and he was correct. But apparently that was impossible, because it would violate Maxwell's equations, where any ripples in the fields have to move at the speed of light, they cannot stand still.

An important consequence of Maxwell's equations, is that they demonstrate how fluctuating electric and magnetic fields, propagate at a constant speed c in a vacuum, which is the second of the two postulates of Einstein's Special Relativity. Consequently, that second postulate, was it already known at the time, openly stated by James Clerk Maxwell? While the first postulate was simply Galileo Galilei's Relativity. So, what did Einstein discovered exactly, with Special Relativity?

And I'm not sure what the problem he was attempting to fix was, exactly. These ripples in the fields do not stand still, they are moving at the speed of light. And Einstein would have as well, had he been running alongside the beam of light. Thus, this is what he would have seen, the explosion of the Sun frozen in time.

This is how Einstein resolved the problem, he concluded that even if he were to run at the speed of light alongside a light beam, that beam would speed away in front of him at the speed of light. Therefore, the speed of light was constant at all times for anyone anywhere, no matter their speed or direction of motion, and no matter the speed or direction of motion of the light source.

That is quite a statement, especially when this light is now made of clusters of electrons, instead of energy photons, and that they can be compared to asteroids hurtling into space. What applies to one, must equally apply to the other. Einstein's statement cannot be correct, even if somehow relative motion, could make it that everyone in different locations, could all more or less measure the speed of light as being c, which I doubt anyway.

After a few thought experiments involving moving trains, Einstein's solution was that observers in relative motion, experience time differently. It's perfectly possible for two events to happen simultaneously, from the perspective of one observer, yet happen at different times from the perspective of another. And both observers would be correct. Einstein concluded that it was the simultaneity of the event that was relative, meaning that it was time that was relative, instead of the speed of light itself.

And suddenly, this constancy of the speed of light, and time becoming relative, and a dimension linked to spacetime, would only work if mass and energy were also relative to the point of view, and were interchangeable through $E=mc^2$. The concept of relative mass was born, where moving objects can become so massive, when travelling at near the speed of light, that it would require an infinite amount of energy to accelerate them any further.

Which seems impossible, when considering relative motion. Something can only become more massive, from the point of view of someone else elsewhere, with a different mass measurement for another elsewhere, but nothing is really becoming more massive within its own frame of reference.

And we know now that $E=mc^2$ is not a magical equation of conversion of matter into energy, and vice versa. It is simply the equation for kinetic energy for moving objects. Because light is not energy, it is matter moving into space. And with relative motion, and the geometry of expansion, light can gather speed and have a real force of impact when it hits something. How could $E=mc^2$ be used to justify the relativity of mass and energy, just because for everything to work, time had to be relative and the speed of light constant?

From there on, time was a dimension all of its own, the fourth dimension, alongside length, width and height. And with all of this, came the erroneous notion that the speed of light is an ultimate speed limit imposed upon us. From one big mistake, many others followed, in order to make it all work. And it continues to this day, with dark matter, dark energy, gravitational waves, graviton particles, black holes, Big Bang and singularities, none of which will ever be found, even if somehow it is claimed to have been found. Beware. I am discussing all these concepts in the articles I have written, which are included in the last chapter of this book.

When we stop for one second to consider all this, it is obvious that it is counterintuitive, and goes against common sense. Consequently, let's forget all of this, and see what would really happen, if we were to move back and forth at different speeds, including faster than light, to observe the event of the explosion of our Sun from afar.

First of all, we are not talking about one little wave of light or energy going across space, that we need to catch up with. We are talking about a massive sphere, spanning several galaxies near us, as far as they can see the light of our Sun, completely filled in all directions of all the clusters of electrons of light emanating from our Sun. And depending on where we are in that sphere, we can see through time. I did not say travel in time, mind you.

We can see through time, as long as we can travel faster than light, to see the event of the explosion of the Sun, before and after the explosion. We then simply see the clusters of electrons of light at that particular location, at that particular time, which will show us if the star has exploded already, or not, or even freeze the explosion if we are moving away from the Sun at the speed of light.

What is important to understand here, is that it is not all of the light around us that would freeze, it is only the light from the event of the exploding Sun. There are always multiple beams of light, coming to us from every single source of light out there that we can see. And also, the light reflected from the surfaces of all the things around us, which is actually how we can see everything that surrounds us.

If suddenly our Sun were to explode and go supernova, on that spaceship going at twice the speed of light towards the Sun, they would see this explosion happen in fast forward. They would see it happening at more than twice the speed of light, maybe three times the speed of light, when adding the speed of light to their own speed. If they were to run away from it at the speed of light, when they look back, they would see the explosion frozen in time. And if they were to escape at twice the speed of light, they would see our Sun existing in slow motion before it goes supernova.

That seems to be common sense. But as I indicated above, the speed of light is subject to relative motion, it varies and it accelerates. Each moment we would have to measure the exact speed of light using the Atomic and Subatomic Expansion Equation, which is the same as calculating the gravity between expanding subatomic particles and the spaceship. The explosion of the Sun itself would affect the speed of light at that point. And all the planets that light would encounter on its way out of the solar system, would as well.

As to how we would see that spaceship, while it is going faster than light, or even just a fraction of the speed of light, but still a considerable speed, I have some empirical evidence to offer, real time observation. I saw a UFO once in my region of Saguenay-Lac-St-Jean, in the North of Québec, right after a power station nearby went bust, highlighting in bright yellow, every electric cable in the streets. The UFO went horizontally a distance away, from right to left, at a speed I believe was faster than light, although I cannot be sure.

What I saw was a multiplication of that same ball of light, until an entire line on the horizon was filled with dozens of that same ship. It started to fade from right to left as it moved, until I saw just one light where it had stopped. It is a safe bet that something moving faster than light, or even just at high speed, would appear to us to multiply depending on its speed, and no one could argue that there is still just one UFO, and not several of them.

An interesting point about this event, is that beyond a certain speed, we see a multiplication of the observed object. So possibly, when our spaceship is going faster than light towards the Sun, they would see multiple Suns exploding, as they come across the various clusters of electrons of light in space, at high speed. It would be like seeing the event in fast forward, but it may take the form of multiple super imposed images of the event. Which would make sense, since different clusters of electrons would enter their eyes and camera lenses, all the time, and sometimes all at the same time.

And for this entire feat of high speed on such a short distance, there was nothing like the UFO or light contracting in any way, it remained at a constant size the whole time, contradicting Lorentz contraction. Possibly the UFO was not going anywhere near the speed of light, but it is something to think about.

Especially when comes the time to try to imagine what we would see, when objects are going faster than light, and when we eventually go faster than the speed of light ourselves. None of this would make any sense within Einstein's ideas, since nothing can go faster than light in Special Relativity, while the entire universe has to change and adapt for this concept to remain true.

Concerning how long light would take to arrive on Earth, from two different stars exactly 10 light years away each, depending on how much matter there were in their path, light might arrive 10 years later from one star, and 11 years later from the other. How much matter there is between us and the stars, would show in the light redshift measurement. Redshifts are no indication of distance, velocity or direction of motion, of astronomical bodies in the universe. Instead, redshifts are indications of how much stuff light must go through in its travel to us, as you can read in my article included in the last chapter of this book: *Dark-Matter, Dark-Energy and the Big-Bang All Finally Resolved.*

In practical terms, there is always stuff in space in the light's path, which makes it extremely difficult to measure accurately, as it is never constant. And with motion, which is relative, then that speed of light is further affected, depending on your frame of reference or point of observation.

No matter how long light might take to arrive to us, and everything that can slow it down or accelerate it in its travel, there is this idea that once it is here near us, it could be near constant. Through a string of adjustments normalising its speed, due to relative motion, the geometry of expansion, and so on.

Nonetheless, if you remember the example of the spaceship going at twice the speed of light, with its beacon of light, their speed of light c from the beacon, and our own value of that speed of light from that same source, could not be equal. Even if both the people on the ship and us here on Earth, were to measure the speed of light to be around c. The speed of light might be c for them, but from our point of view, they are going at twice the speed of light. In effect, the speed of light is relative to the point of view, to the frame of reference.

And of course, this entire discussion above does not take into account that everything is actually located here with us, as there is no distance in this universe, and not even light covers any distance. I had to simplify the discussion in terms we are all familiar with, but the switch is not that difficult to visualise. There is still that sphere surrounding each of us, where we see everything surrounding us, but it is more like a smaller bubble around us, within our own brain, than a huge sphere spanning entire galaxies out there in the distance.

Whenever we see light coming from somewhere, the electron clusters are not travelling to us, they are simply expanding towards us, just like asteroids would, if they were to move from somewhere in space towards us. And the same if they are moving away from us, they are simply shrinking away to the periphery of our bubble reality, unique to each of us. We are living in an egocentric universe, where we are the centre of a universe of our own making. I call it a psychophysical universe, but it could be argued that it is entirely psychological.

EIGHT

Main points of the Theory of Universal Relativity

8.1 Matter expands unvaryingly at the same constant rate in all realms and dimensional worlds, independently of relative motion or any frame of reference (first type of expansion)

The expansion rate of neutrinos, electrons, atoms, molecules, planets, suns, solar systems and galaxies, is the same: X_A = 0.00000077 $/s^2$ (or 7.7 x 10^{-7} $/s^2$). This inner expansion rate is constant across the universe, other realms and other dimensional worlds, and is independent from relative motion or any frame of reference.

As a result, we have distinctive scales of universes in the very small (subatomic and atomic matter) and the very large (astronomical objects), that never reduce in size. They simply constantly expand, along with their orbits, at the same constant rate. The speed at which all particles and astronomical objects orbit everything, must constantly increase as well. Which means that the frequency or the vibrational rate of matter, constantly increases steadily. However, it does not increase in an exponential acceleration.

That electrons, atoms and molecules, can grow larger or smaller than our usual ones, vibrating slower or faster while changing the density of matter, creating other planes of existence and other dimensional worlds, does not change the universal expansion rate of matter. If it did, it would not take very long, before these other worlds grew or shrank disproportionally compared to ours.

And from the evidence we have so far, of the existence of such worlds, indicates that they share our space in normal physical worlds, that exist in parallel to ours, or intertwine with our reality, just like gases do in the air. From the literature and witness statements, people have easily switched from one world to the other. It would not be possible, if the other worlds were expanding any slower or any faster than ours. Therefore, the same matter expanding at the same expansion rate, composes all worlds.

8.2 Gravity is due to the expansion of matter, and expansion causes motion

All five types of expansion and contraction of matter cause motion. The first type is the inner expansion of atoms and electrons, which reduces the distance between atomic and subatomic objects, in two types of distance decrease. This is gravity, as stated in Mark McCutcheon's Atomic Expansion Theory. It is calculated using the Atomic and Subatomic Expansion Equation, which replaces Newton's Law of Universal Gravitation, and Einstein's General Relativity.

The absolute decrease in distance is due to the own internal expansion of the objects. There is also a further relative decrease in distance, due to the fact that space between the objects does not expand along with the objects, since space does not exist as a concrete thing made of expanding matter. From one moment to the next, the radii of the objects are getting longer, compared to the distance which remains the same, causing the acceleration of the distance reduction. This is an effect of relative motion and the geometry of expansion.

These expansion rates of atoms and electrons are invariant across all frames of reference, as stated in the first point above. This expansion causes a natural motion which leads to the natural orbit effect, as distance between all objects reduces due to expansion. If, from the first perspective behind the scenes, the objects are moving fast enough in more or less straight lines to avoid collision, from the third perspective behind the scenes, they will orbit each other in enlarging spirals, compensating for the constant expansion of all objects.

8.3 Motion is the relative expansion and contraction of matter (fifth type of expansion)

The fifth type of expansion and contraction of matter, concerns the relative motion of objects, and is relative to the viewpoint. It is molecules, atoms, electrons and neutrinos, including light, expanding and contracting with motion, depending on the frame of reference. With the relativity of motion, there is only the relative acceleration or deceleration of the rate at which matter expands or shrinks, compared to us. There is no more distance covered by moving objects.

Nothing really ever moves in Universal Relativity, or gather any kind of velocity. Everything is here, and it simply relatively expands or reduces in size, compared from one point of view to another. To visualise this, it is best to think in terms of how would a virtual reality, or a simulated world, be constructed in 3D on a computer monitor. To create the illusion of movement, the objects would expand or reduce in size, compared with other objects on the screen, and they would not really cover a distance in space.

It is still perceived as motion by us, and it could be said, almost as an optical illusion. When someone walks away from us, in the direction of their car, that person only shrinks away from us.

However, from the point of view of the people in the car, that person is expanding while coming closer to the car. This is relative motion.

8.4 The laws of physics and our variables are invariant within all frames of reference taken independently. With relative motion, our *visual* variables become relative and changing, when compared from one frame of reference to another, but it is only an optical illusion

Within all frames of reference taken independently, no matter our motion or acceleration, none of our variables in physics change. Time, distance, size, mass, energy, velocity and the speed of light, although the speed of light is not constant, are all unchanging and non-relative values. This is the invariance of the laws of physics within any frame of reference, as stated by Galileo, which is also the first postulate of Special Relativity.

With relative motion, when comparing one point of view to another, all *visual* variables in physics like distance, size, velocity and the speed of light, become relative values. These values for one frame of reference, would differ from the point of view of observers in other locations.

Motion is relative between yourself and everything else. You cannot tell what is moving, meaning expanding or reducing in size toward you or away from you, and at what speed, meaning at which expansion rate. And different people in different places, won't see the same thing at the same time, from the same event happening in one specific place.

The relativity of motion brings optical illusions, because the speed of light takes time to travel. The speed of light is also subject to relative motion, it slows down and disperses as it travels, and is accelerated by slingshot effects and gravity, a changing distance between expanding objects, obeying the laws of the geometry of expansion.

There is nothing preventing us from going faster than the

speed of light. And since the speed of light is relative, while travelling faster than light, you would still see just fine. It is just a question of the clusters of electrons within light, to travel from any source to you, and it does not matter at what speed you are going at. Other sources of light around you, are not connected to any other source of light out there. Each are independent from each other, as they are all different clusters of electrons travelling in space at different velocities. Your measurement of any speed of light, could easily be similar to our normal speed of light here on Earth, even although you would be going at several times the speed of light.

And then, your speed of light would not be the same as the one of someone at rest, compared to you. As a consequence, the speed of light is not constant in all frames of reference. It varies, once you start comparing it from one frame of reference to another, as the speed of light is relative to the point of view.

As for our other variables in physics that are not visual, like time, mass and energy, they are no longer relative. They cannot be affected by the time it takes for light to arrive to us, which distorts reality and what we see in the universe. There are no longer such concepts as space-time and mass-energy as stated by Einstein.

8.5 The speed of light varies and is relative, the configuration of the universe is relative

The speed of light varies and is relative, it is no longer constant in all frames of reference, always going at c, at approximately 300,000 kilometres per second, or 186,000 miles per second. The speed of light is only relative in the sense that motion is relative, and light is not a special entity that requires its own special laws of physics.

Light is made of clusters of electrons, that are also simply subject to the relativity of motion. You cannot tell the speed of light, until you identify a specific frame of reference from which to measure it from. And then, you still cannot tell really what the speed of light is, because you don't know your own speed and direction of motion to begin with, neither the real speed of light or its direction of motion.

In all cases though, and this is the most important part, the speed of light should be calculated as gravity using the Atomic and Subatomic Expansion Equation, and considered as the changing distance between two expanding objects. We must consider the initial speed of the light from the source, as clusters of electrons in the crossover effect, are launched freely into space at quite a speed out of the subatomic world. Gravity is an acceleration, hence the speed of light, or the reduction in distance between its origin and its destination, accelerates.

With the distortions of the limitation at which light arrives to us, we cannot trust anything we see in the universe. For example, the gravitational lensing effect, where the path of the light is greatly affected by the geometry of expansion, and accelerated, as it passes by galaxies, stars and planets. Such effects radically changes where objects are truly located in space, and their apparent speed and direction of motion.

The configuration of the universe is equally relative. With the use of transformation equations, we can then assess the real value of the speed of light, from one frame of reference to another, the deviations of the beam of light, where objects are really located in space, and calculate their true speed and direction of motion. However, in light of everything I stated, our actual transformation equations, are hardly adequate to measure these at this time. At least time ticks more or less the same everywhere, in our own dimension that is.

8.6 Time is neither relative nor a dimension, time is constant everywhere

Time is not relative, nor a dimension, as it is not the same as size, length and distance. It ticks more or less the same everywhere, there is no reason not to, because time cannot be affected by what we see, or distorted by the time light takes to travel. The term *optical illusion* says it, any relative value can only be one we can *observe* being distorted, by relative motion and the time that it takes for light to travel to us.

Time is just a practical convention. It ticks the same everywhere, although under varying conditions and other mechanical stress, it could appear that time runs at slightly different rates. However, this is not due to any relativistic effect as Einstein proposed, and time could never change that significantly.

The spacetime concept of Einstein is now obsolete. Time is no longer a dimension linked to length, width and height, although this model might still describe what we see through the relativity of motion, to a certain extent. But it is just an optical illusion, while time is no longer relative.

There are many proofs that time is relative, these proofs are simple misinterpretations. Time could be running at different rates with altitude and acceleration, simply due to mechanical stress and other variables, like temperature, humidity, radiation, magnetic fields and the accuracy of our technology.

Lorentz was saying that with acceleration, the spacing between the particles simply reduces, the object becomes denser and contracts, this led to his transformation equations. This is when Einstein thought the relative effect on reality was real, and decided that time was also affected.

Possibly, with acceleration or altitude, the spacing between particles might change slightly, but only due to mechanical stress and other factors, and it might affect the running of clocks. However, this is not the result of a relativistic effect of spacetime, and it must be incorrect that at near the speed of light, an entire spaceship could in reality shrink to almost nothing, or a simple dot. From our point of view, it might visually appear to be so, but it would only be a visual relativistic effect, due to relative motion and the time it takes for light to travel.

8.7 Interdimensional time is relative, but not as Einstein proposed, and other dimensions are defined by the frequency at which neutrinos and electrons are orbiting within and between atoms

Interdimensional time could be considered relative. Meaning the rate at which time goes, changes depending on which dimensional or density world you are in, which entirely depends on how fast your particles are orbiting and vibrating.

The interdimensional frequency or resonance of your matter, defines your time rate and density of matter, and in which dimensional or density world you are in. If you are vibrating faster, and your particles are larger than normal and more spaced out, then time could be perceived to go much faster than normal. Years could go by, in higher dimensions, within just a few hours on Earth.

This type of relativity to time, is unrelated to what Einstein proposed, that the time rate changes with acceleration and deceleration in Special Relativity, and with altitude and near significant gravity fields, such as the Sun, in General Relativity. The relativity of interdimensional time, is entirely due to the vibrational rate of matter.

8.8 Mass and energy are not relative, although they are subject to relative motion

As demonstrated in *The Final Theory,* in the section called *$E=mc^2$: What is Energy-Matter Conversion?,* the equation $E=mc^2$ is derived from the equations for the kinetic energy of moving objects. Light is made of clusters of electrons, which are moving objects. In other words, $E=mc^2$ is the equation for the kinetic energy, or momentum, of freed electrons expanding outside the atomic and subatomic realms. And of course, there is an equivalence with how many electrons were contained within the mass of the atom, from where they were freed. Protons and neutrons are also entirely made of electrons.

$E=mc^2$ is an equivalence between electrons in two different forms: electrons freely expanding outside the atom for the E or energy, and electrons within the atom for the m or mass, multiplied by the speed of light squared. Where the speed of light c corresponds, more or less, to the speed at which the electrons are freely launched into space, during the crossover effect. The square is an indication of the acceleration of these expanding electron clusters, once freed from the atoms, due to gravity. Gravity being the distance reduction between everything, due to the inner expansion of all matter.

Mass and energy are not relative. The momentum for light, or any mass, or the force of impact, will change due to relative motion and the geometry of expansion, but it will change in the same amount for all observers in different locations. The changing kinetic energy of that mass, will be the same for everyone, no matter where they will see it from, and when. In that sense, mass and energy are not relative to the point of view.

Like time, they are not visual variables in our physics. For a value to be relative, people in different places, must measure that value to be different from their point of view, than for others elsewhere. This could only be true of visual values, that can be distorted by the time it takes for light to travel. An object cannot be more or less massive or energetic, depending on the point of view. An object would simply always have the same mass or energy content, to everyone in any frame of reference.

No matter how much kinetic energy we measure a mass to have, from one frame of reference to another, it does not change the final momentum, or force of impact, that the mass will have when it hits something. And this is true, even if that mass goes through a slingshot effect, or enters into an orbit, or is accelerated through gravity.

To clinch the argument, if you read online about the concept of relativistic mass, you will see that there is controversy in the scientific world, about if mass is actually relative. Some people in science, ignore the concept of relativistic mass altogether, for practical reasons. But also, because it leads to impossible scenarios, where, for example, an electron could be more massive, or have more energy, than an entire atom.

You will also find, that it is more the entire concept of mass-energy taken as a whole, which is considered to be relative, where mass would remain the same no matter the frame of reference and motion, and only the energy level would be relative and changing. But, is it just changing because of relative motion and the geometry of expansion, while remaining the same for everyone everywhere?

As mentioned just above, an object or system cannot have more or less energy, just because of where an observer is trying to measure it from. It has a certain mass, a certain amount of energy; and all the effects that can lead to the amount of kinetic energy increasing, cannot depend or change for observers in different locations.

NINE

Third type of growth and shrinkage of matter - Other dimensional worlds or world densities

9.1 Third type of growth and shrinkage of matter - Interdimensional density of matter, when the electrons, atoms and molecules' vibrational rate, size and spacing vary

The third type of growth and shrinkage of matter, is when the neutrinos are orbiting faster or slower, around and between the electrons, enlarging or reducing their orbits. This is the cause of the change in size of the electrons, which are essentially solar systems at the smaller scale, with the planets being neutrinos, and the sun or suns being super-neutrinos.

This process is identical to the crossover effect, in the second type of growth and shrinkage of matter, when the electrons grow and leave the atoms to become anything that is in the electromagnetic spectrum. The difference is, the change in size of the electrons is happening within the atoms. When electrons grow or shrink in size within the atoms, it affects the size of the atoms and of the molecules, their level of vibration or excitement, and the overall spacing between all these particles.

This has never been obvious or known to us before, in our

science, because when matter changes density in that manner, it moves outside our range of visible light. But more than that, it is like this matter is no longer part of our reality, since we can no longer interact with it. From our point of view, matter in the lower dimensions move in slow motion, while in the higher dimensions matter moves in fast forward. Which might begin to explain why that matter disappears and is out of our reach.

If the change is not too significant, or the conditions keeping these worlds apart break down for whatever reason, we can still see and interact with matter vibrating slower or faster. But the consistency of that matter is suddenly different than what we are used to. Rocks and metals are unexpectedly malleable, sometimes intangible, they can levitate easily as if lighter or less dense than air, and no doubt this is how they built the pyramids and Stonehenge.

Beings from higher dimensions, or from the future, glow, simply because their neutrinos, electrons, atoms and molecules vibrate faster. And from the lower dimensions, everything seems very heavy, even in the atmosphere, and beings are sometimes dark and shadowy, instead of substantial. However, if they cross over completely, they can be as solid as anything.

The density, when it comes to other dimensional worlds, or world densities, is a different type of density of matter, than the one we find in our books of physics today, although they are closely related. All dimensions share the same space, the same Earth, just like the gases in the air surround us. All these worlds use the same particles that we do, all expanding at the exact same inner expansion rate. And they too have solids, liquids, gases and plasmas in their own dimension.

Usually, a change in density is simply a change in the spacing between the molecules, which turns a solid into a liquid or a gas. Indeed, more electrons added to the system through heat, or pressure being applied, will cause the electrons to orbit faster, enlarging their orbits between molecules, but not between or within the atoms, so atoms and molecules don't change size. This excites the atoms and the molecules, causing the molecules to become more spaced out. This is all within our own third dimensional world, where electrons, atoms and molecules never proportionally grow or shrink, while still expanding at a constant rate, from the first type of expansion of matter.

Although there is the same type of change in density in all dimensions, of solids, liquids and gases, each dimension has its own specific vibrational rate, size and spacing for the electrons, atoms and molecules. These are the only difference. Something does not need to vibrate that much faster before it disappears from view, since some insects can easily do so when in flight.

This is why it has always been confusing, and why we never understood what the spiritual literature meant, by another density world. It is also a change in density, in a different manner, when the particles become larger, as well as more spaced out, while the neutrinos are orbiting and vibrating faster, causing the electrons to grow within the atoms.

It is a change in the size of the particles and the spacing between them, which is a result of the change in the vibrational rate of matter, beginning with the neutrinos causing the growth or shrinkage of the electrons within the atoms. Otherwise, a simple change in the vibrational rate of matter, of the electrons without growth or shrinkage, solely leads to a change in the state of matter, and not to matter moving to another dimension.

We must bear in mind that, matter moving to another dimension, appears to bring new states of matter unfamiliar to us. However, it might simply be how we perceive it from within the third density world. From within their own dimension, other beings might perceive it as normal states of matter, similar to ours.

9.2 Proving Earth used to be smaller and denser, while weight and gravity are increasing with time

If we could prove the third type of growth and shrinkage of matter, we would essentially confirm the existence of other dimensional worlds. In concept, since it is similar, it would also validate the second type of growth and shrinkage of matter. Except of course that, in the second type, the electrons grow and shrink outside the atoms instead of inside. This is no small test, and the proof certainly would be revolutionary.

At this present time, it is not obvious how it could be properly tested. Except perhaps, through proving that the Earth used to be smaller than it is today, albeit denser, and that weight and gravity not only used to be less, they both still increase with time.

It would explain why gigantic animals, such as dinosaurs and pterodactyls, came to be in the first place, and had no issue moving or flying around, on a smaller planet without oceans, mainly covered by all the continents. This was discussed already in section *2.7 How to move to other dimensions?,* where there are links to Neal Adams' animated simulations, showing that all continents used to fit perfectly together into a smaller sphere.

Let's begin with this quote from section *1.2c The third type of growth and shrinkage of matter:*

> "The process causing the expansion of all matter behind the scenes, will be discussed shortly, in the section about gravity and orbits. But for now, most of the natural acceleration within orbits, and the enlargement of orbital rings, does not affect the interdimensional density of matter, or the size of particles and objects, within our resulting reality. But a tiny part of this process is, and in time, particles and objects orbit faster, and grow slightly more, even in our resulting reality.
>
> "Consequently, as time goes by, all particles orbit slightly faster, all orbits enlarge slightly more, all particles and objects

> become larger, and the time rate marginally accelerates, as we constantly move into higher dimensions. Which makes any time and place in history, its very own dimension, where matter vibrates at a unique frequency, or unique resonance."

The test discussed below, is not exactly about proving directly the third type of growth and shrinkage of matter. Instead, it is about proving that a tiny part of the invisible process, of the expansion of matter, leads to an actual and real expansion of matter, in our resulting reality.

If that is true, then it proves the third type of growth and shrinkage of matter. But if it is not true, it does not disprove it. Because the third type of growth and shrinkage of matter could be true, even if the tiny expansion of matter in time, within our resulting reality, is untrue.

It should be extremely simple to test that weight and gravity, not only used to be less than they are today, but they both still increase with time. Right? If we drop a mass of 1 kg from a height of one metre, and weigh it once it is at rest, carefully measure the results, and repeat the experiment every year, we will notice that weight and gravity increase with time.

Why? Because particles and planets are increasing slightly in speed all the time, causing the orbit and the size of all the particles, and of all the astronomical bodies, to enlarge, in an expansion of all objects that is visible and can be measured.

Consequently, the Earth and everything else grow in time, enough to cause the continental drift, meaning the continents are constantly slightly moving away from each other. Moreover, the distance between objects reduces faster as they grow, which is gravity increasing. And the force required to move, accelerate, or hold an object at rest against gravity, meaning its weight, would also increase with time.

But if it is that simple, why have we not noticed it before? Countless experiments have been done. Surely it is verified, that a mass of 1 kg must always fall at the same rate, and its weight, in a same location, must always be the same? As it turns out, it is not so simple to prove this point after all.

First, it was assessed previously in sections *1.2a The first type of expansion and contraction of matter,* and *3.4 Gravity and the formation of galaxies,* that objects don't all fall at the same rate. It was stated that our measurements of falling objects on Earth, are too limited to even notice a difference, between a falling bowling ball and a feather in a vacuum, or a massive aircraft carrier and a marble, dropped from high up. They do appear to fall at the same rate, when theoretically, whether we use Newton, Einstein, or Atomic Expansion Theory, they should not. In all three theories of gravity, larger or more massive objects should reach the ground first.

And if the experiment is not performed in a vacuum, terminal velocity could skew the results. At some point, and not all at the same point, the falling objects would stop accelerating, until they reach the ground, because of the objects' buoyancy and the friction with the air.

Therefore, from how high up must we drop two objects, of different size and mass, in a vacuum, to finally notice a difference? Excellent question, which requires an answer before we perform any test or experiment.

Another major issue is that, two different objects with a mass of 1 kg each, cannot reliably be trusted to be 1 kg at all, or to have the same mass, unless we compare them side by side. Which must bring discrepancies between different experiments, performed at different times and locations, using different masses and instruments.

Furthermore, it is said that a mass of 1 kg is always the same. It is true that the number of electrons, atoms and molecules of a 1 kg object, should theoretically always remain the same, even when the object becomes less dense and enlarges. But how do we measure or calculate this mass?

We establish the mass of an object using its weight, also the force required to move that object – or its resistance to acceleration – and the force of gravity. All of which change, depending on the location of the object on Earth or in space. An object of a certain mass will not weight the same at the equator, in London, or on the Moon, since the force of gravity is not the same in all locations.

Plus, gravity is different depending on if the object is free falling, where only the size of the objects matters, or if it stays on the ground, where the density of matter also affects gravity on the ground. Gravity on the Moon is not uniform. While it is one sixth the gravity of the Earth on the near side, it could be up to one third the gravity of the Earth on the other side, where the composition of the ground is less dense than on the near side, where the centre of gravity is located.

And finally, if all objects in the universe become less dense with time, we might not notice any difference in our measurements. The interdimensional density of matter is not the same as the traditional density of matter. It is not solely the space between the molecules which changes, it is also the size of all the particles, as well as the spacing between them, which change with the acceleration of all particles.

To make matters worse, the teams in charge of establishing the International System of Units, works very hard to ensure that all units remain exactly the same, year after year. Of course, a kilogram, a second and a metre should always remain the same, otherwise how could we compare accurately our measurements?

But what if our kilogram, second and metre really change every year? Means would have to be found to ensure that they don't, that they are constantly adjusted to correct any discrepancy. As a result, we might believe that gravity and weight do not increase with time, when in fact they do.

There is a string of constants in nature, which we use to measure accurately most of our units. But then again, how can we be sure that these constants are truly always constant, especially in relative terms, and that this does not skew the end results?

For example, we could not use the size of the Earth to establish the length of a metre, as we used to in the past, since both the Earth and the metre stick may be growing. I heard that a metre does not exactly measure the same, neither is the speed of light, from one year to another. This is not surprising, because today we are using the speed of light, to accurately calculate the length of a metre. But this is a mistake, since the speed of light is not constant, as discussed in the first three sections of chapter 5.

And how do we measure the speed of light? We use the second and the length of a metre. It is a vicious circle, we use variables to measure variables, that were originally measured, by what we are measuring. So, if all these variables increase in time, or change, we will fail to notice, as they are constantly self-readjusting, and self-standardising each other.

Also, we cannot accurately measure gravity using light or light beams. What we see is always relative, because the speed of light is relative, and variable. And now that time and mass are no longer relative, meaning they are constant, there must be a bunch of measurements out there requiring correction.

And then, the second is defined by using the vibrational rate of the caesium 133 atom. But if the atoms, and the electrons within the atoms, are constantly growing, and the electrons are accelerating, and their orbits are enlarging, then it must also affect the measurement of the second.

That is all three units required to measure gravity and weight – the metre, the second and the mass (measured through weight and gravity) – which potentially we have been adjusting and normalising as time goes by, while for all that time, they may really have been changing, without us realising. Therefore, how can we be sure that we are measuring gravity or the speed of light correctly?

In view of this new physics, if the speed of light, and the frequency of a caesium 133 atom, are not constants of nature, then what about the other constants of nature, upon which the measurements of our units depend?

Everything changes in time. Particles, orbits and objects are expanding, both behind the scenes, and in our resulting reality, while we are not even aware. And although we might think that our measurements are always the same, relatively and proportionally speaking they might not.

This might begin to explain why, if the Earth is growing in time, and the density of matter is becoming less, that weight and gravity also increase with time, and that the speed of light is variable, we might have failed to notice. It also shows how it might be difficult, to verify such a state of affair. But it might be within our capabilities to prove it, if we specifically set ourselves to do so.

To prove the third type of growth and shrinkage of matter, and essentially to prove that weight and gravity increase with time, we would need to drop the very same object from a very high height – and I mean at a significant distance – at the very same location, year after year. We would need to use the very same measuring instruments, to measure the fall and the weight on the ground of the object when at rest. And it would have to be done in a total vacuum, or as close to it as it is possible. Maybe it could be tested in space, outside the International Space Station, but I recommend the Moon.

Despite all this, I do wonder if we could prove it. If time and gravity are accelerating, the metre is getting longer, and mass, or at least its weight, is increasing, is it possible that we would still get the same measurements year after year, even although these values increase in real time?

Measurements are, after all, just an equivalence and comparison between quantities and variables. If all variables increase over time, we might get the same relative and proportional results, and be unable to prove any of it, even when it is happening in real time, and that such changes must be visible, and should be measurable.

This said, there might already be some data or proof somewhere, unrelated to measuring falling objects or weighing them, providing an indication that this is true. For example, the evidence that once upon a time, all continents were glued together to form one large continent called Pangea, which led to the idea of continental drift.

And now, it seems that all continents once formed a perfect smaller sphere, as if the Earth was much smaller in the past. This is such an indication that these theories might be correct. This is not to say that the plate tectonics theory, to explain the continental drift, is wrong. However, what drives these plates to move and cause earthquakes, could be the growth of the planet.

Let's hope people will start looking into this and find other evidence.

9.3 Interdimensional time and higher dimensions - Time is relative to the density and vibrational rate of matter

What is meant exactly, in the New Age, spiritual and esoteric literature, when they say that time does not exist? Do they really experience their entire existence all at once, in a simultaneous timeline, or spacious present? Like a computer programme sitting on a server somewhere, not being played, containing all the possible outcomes of its playthrough? Or multiple instances of the same game opened on a computer, all being played at the same time?

Is it because they can switch to the past, present or future, just like that, all the time, just by thinking about it, and bang they're there? Hence there is no possible linearity to time in their mind? Or every time they stop somewhere, time runs at radically different rates, so keeping track becomes secondary? How do they measure being babies compared with being old, or that an action must precede a reaction? Or even, the simple concept of before and after?

Perhaps the ultraterrestrials live in a world where reactions might precede actions, or that results might happen without even an initial action, or that they experience being babies and old pensioners all at the same time, who knows? But maybe it is just semantic, and that the way they live and experience life, led to a different kind of measurement of time, or of these concepts just mentioned.

You certainly would not keep a watch, if you could constantly travel in time, and everywhere had a different time rate. Or if you kind of perceived living your entire existence all at once, in a fluctuating timeline. Simply because, you are fully aware of both your future and your past at all times, along with the constant changes to both.

Or maybe they only ever have a present, while living in different bubble realities. Where within one bubble, a certain past already exists as a given, like in dreams, which might change from bubble to bubble. I'm not sure exactly how that works out for them, or if this is really what is going on here.

You think about a certain moment in time, and here you are there. Like if any specific event was actually just a bubble reality, or a dream, that can change all the time just by thinking about it, and that you can experience at any time. It would be like being connected to the programme running on a holodeck, where you could recreate and relive any moment of your existence at any time. The programme would be monitoring your thoughts and changing everything as you go along. Except, your brain or consciousness would create and change everything, not any external programme or computer.

There is something that gives us an idea of what that might be like, although it is just sci-fi, the double first episode of Star Trek Deep Space Nine, where the wormhole aliens are trying to understand the concept of linear time. Commander Sisko keeps coming back to the solar system Wolf 359, about 7.9 light years away from Earth, where his wife died at the hands of the Borg. And the Prophets are telling him something like: "But you live here as well, at the same time as all the other moments in time that you are showing us".

What kind of brain, processor and memory type, they must have as their mind, in order not to understand the distinction, that these are different events on a linear timeline, and what they saw were just memories? And what kind of existence are they really living, compared to us? Star Trek Deep Space Nine does not really tell us, except that the concept of time is meaningless there.

Maybe this kind of life is more suited or appropriate to eternal beings, without a physical body as dense as ours, where we are born, fall in love, get married, have a kid, fall out of love, divorce, and then die. Not necessarily in that order, with perhaps a few more marriages, kids and divorces, but usually in a linear and chronological fashion.

You might have noticed what I have just said, that every single place has its own time rate. When earlier, I stated that time ticks the same everywhere, and Einstein's relativity of time was incorrect. But of course, now I am talking about interdimensional time, where time is still relative, depending on the dimension or plane of existence you are in. And the reason time is relative there, has nothing to do with Einstein's relativity of time.

Within our own reality, generally limited to the third world density, or third dimension, time ticks the same everywhere, more or less. Although, technically, every single place in the universe is its own dimension or plane of existence, its own bubble reality, with its own time rate. But if they are all of a similar density of matter, and vibrate relatively all at the same rate, then they can all be considered to be part of the same dimension, or world density, where time should relatively tick the same everywhere, within that overall dimension.

Reading the book of Dolores Cannon called *The Custodians,* it appears most phenomena involving the loss of time, in alien abduction cases, could be explained by the fact that the abductee was either not conscious, in an altered mind state, or his or her memories were somehow changed or wiped out. They don't always realise where they were for the time they were abducted, and not aware of the passage of time. Going to other dimensions does that to our memory, it is like trying to remember dreams.

Plus, instances of condensed time, where suddenly people appear to have arrived faster at their destination, can be explained by technology used by these beings. They appear capable of dematerialising a car and its occupants, and instantly re-materialising them elsewhere, closer to their destination, giving the impression that somehow time was gained.

These examples attempt to explain, how time did not need to change rate in these particular cases. But as soon as you move into a higher or lower dimension, time would be affected. I take it that if you can move a car and its occupants into a higher dimension, and back to our dimension, you would indeed dematerialise and re-materialise them elsewhere, from our point of view, but it is just that they moved dimensions. And since the time rate would accelerate significantly, while being into higher dimensions, they would indeed gain time and arrive faster at their destination.

Abductees brought to higher dimensions, could stay there a very long time, before they are returned to our dimension, where almost no time at all would have passed. Because while in the higher dimensions, all their particles were vibrating faster, and they lived in fast forward.

Have you ever noticed, how Presidents and Prime Ministers in the world, appear to age extremely fast while in power? You can search online for photos of before and after their time in power, and it is unbelievable. The most striking examples that come to mind, are Barack Obama and Tony Blair.

Even Gordon Brown, although he was only in power for three years, was ageing and degenerating before our eyes. And this was not the case while he was Chancellor of the Exchequer for ten years, it only happened once he was Prime Minister. And yet, both jobs must be extremely stressful. It is likely these leaders spent some time in higher dimensions, while in power, or were sharing their body with some entity from another dimensional world, and they aged faster as a result. And the question is, were they aware of it at the time?

My guess is that they were aware of it, and that some of our leaders are in contact with other alien species and/or ultraterrestrials. There is a book available online for free on that topic, which I highly recommend, where it can be seen how they were in contact with our leaders in the past. It is called *Genesis For The New Space Age* by John B. Leith. It can also be bought under the new title *Genesis of the Space Race: The Inner Earth and the Extra Terrestrials.*

I read a lot of books about the hollow Earth, and I now believe that it exists, but perhaps only in higher dimensions. And just by going there inside the Earth, we would naturally move to higher dimensions. As to how or why exactly, I'm not sure. I will say this though, if the Earth, moons, planets and suns are hollow, even in our dimension, and there are strong indications that they are, it is clear that the expansion of matter over time, would explain this better than any other theory out there.

I've been in contact with Rodney M. Cluff, the author of *World Top Secret: Our Earth IS Hollow!: The Scientific, Scriptural and Historical Evidence that Our Earth Is Hollow!* I do hope he gets a chance to mount his expedition, to reach the hollow Earth through the holes at the poles. In his book, he even developed his own theory of everything.

And as for any ultraterrestrial's threat to humanity, I believe it is unlikely to come from the higher dimensions. However, there could be a threat coming from the lower dimensions. Entities closer to animals than humans, but intelligent enough to take over the planet. We could say their plan is coming along nicely. I will elaborate further later in this chapter.

To get back to our topic of how ultraterrestrials can snatch us off the Earth, this is apparently accomplished by using a light beam, to accelerate the vibrational rate of atoms and molecules. People caught up in this higher frequency light beam, are not necessarily travelling at the speed of light. I suppose what is meant here, is that our molecules, and the atoms forming them, must be vibrating faster, by getting into contact with this light beam. And seemingly, this is how one moves to another dimension, one where matter is vibrating at a higher frequency. It also appears to be accomplished using technology, as well as through thoughts.

When in a higher dimension, the matter is still there, we just can't see it, because it is of the density of air or a gas, perhaps even lighter. Their particles, such as electrons, atoms and molecules, are slightly larger and more spaced out, while still keeping their normal structure, unlike with gases. And the time rate seems to be affected, but not exactly. When you vibrate faster, you live faster, you can do much more in a same amount of time. The time rate does not change. Put simply, you live and do everything faster. Technically, even here time is not really relative per se, and that is quite an important distinction.

Somehow, when in that light beam, it would seem that our surroundings have frozen, and time would appear to have stopped, or to run very slowly. This is when our molecules are vibrating faster, while normal molecules on Earth are vibrating at a very slow rate, while still being of a normal density. I wonder what kind of light beam this is, but it must be of a high intensity and frequency. Laser light appears closer to that definition, as mentioned in *The Custodians.* However, larger particles, with neutrinos and electrons orbiting much faster everywhere, brings a different kind of electromagnetic spectrum, one more adapted to that different dimension or plane of existence.

Laser is a light amplifier. It stands for light amplification by stimulated emission of radiation. According to Wikipedia, a laser differs from other sources of light, in that it emits light coherently. Spatial coherence allows a laser to be focused to a tight spot. Spatial coherence also allows a laser beam to stay narrow over great distances, it is called collimation. Collimated light, is light whose rays are parallel, and therefore will spread minimally as it expands and propagates.

As for the frequency of light in a laser, it can be any frequency of the second part of the electromagnetic spectrum, from infrared light to gamma rays. I am unsure what size of clusters of electrons these beams would be using for light, to achieve such feats. But most likely, that light is from a higher dimension. It is made of larger than normal electrons and clusters, vibrating faster and more spaced out. This is how they can affect molecules, in order to transport people outside their bedroom, passing right through closed windows and walls, while defying gravity.

John A. Keel, the paranormal investigator and author of *The Mothman Prophecies* and *The Eighth Tower: On Ultraterrestrials and the Superspectrum,* used to call aliens ultraterrestrials, instead of extraterrestrials. Because he believed that they have always lived here, albeit in a higher density world, instead of coming from elsewhere. And several encounters with them, where deceptive attempts at concealing this fact. He also stated that their electromagnetic spectrum seems to be different than ours.

The ultraterrestrials appear to be able to achieve a lot through the use of laser light beams, or bright white light beams. They seem to be used as antigravity devices, and for switching objects or people to higher or lower dimensions. And possibly also to create deceptive virtual realities or holograms, although this could be achieved instead through signals beamed into one's mind, or simple suggestion through telepathy, who knows.

One thing for sure, this is no ordinary light or laser. It must be using a different scale of the electromagnetic spectrum, where the clusters of electrons are somewhat structured differently than ours. The different frequency in light here, refers to how fast the neutrinos within and between the electrons, forming the clusters in light, are orbiting, and the size of their orbits. This is what is meant by a higher vibrational rate of matter. It makes electrons slightly larger, and more spaced out, leading to overall larger clusters of electrons, of a different density and vibrational rate than our normal ones. This is called interdimensional energy.

The frequency here is not the one usually referred to, when talking solely about our own electromagnetic spectrum. Usually, the frequency of light refers instead to the size of the clusters of electrons, while intensity is how many clusters are passing each moment. Nothing about the electrons growing or shrinking in size.

Making matter vibrate faster, changes the time frame, so time could be considered relative in that way. It also changes the density and weight of objects, and then the objects can go through walls, they can levitate easily, as some sort of antigravity. They can probably reach the speed of light easily and even faster, without any of the g-forces we would experience in our space shuttles. Being of the density of air or a gas, no wonder UFOs can float so easily and move so fast, with what seems to be no acceleration whatsoever.

In a book of Dolores Cannon, during a morning session, through hypnosis, they connected to a being in a higher dimension. They decided to break for lunch and re-establish contact in the afternoon. By the time they re-connected, that being was somewhere else entirely, light years away from its previous location, and more than a year had passed since the morning session.

In *The Custodians,* there is a chapter called *The Alien Base inside the Mountain,* a base that would have existed since the beginning of Earth, well before human beings appeared. It would be located somewhere on Earth at a juncture in time:

> "Yes, because there's a juncture of time there. That's why so many of your phenomena take place, is because you're at a juncture in time. The dimensions that come to a point at a juncture in time on the Earth planet and Earth time, cause a twisting and turning so man's perception is altered in such a way that he doesn't really know what happened. He just knows it happened."

Dolores Cannon, *The Custodians*

Advanced humans visit that mountain to be taught, and spending eight hours there is like spending one minute on Earth of our normal time. Hence, there is certainly a way to change our perception of time. But as I showed above, if our molecules and particles suddenly vibrate at a higher frequency, or faster, possibly we could live eight hours in one minute. We would simply gain the ability to do everything much faster, in a same amount of normal time. Everyone else everywhere else would simply *appear* to move in slow motion.

Just like flies vibrating at 250 hertz, compared to us vibrating at 50 to 60 hertz. We are still sharing the same time frame, but flies can achieve four times more than us in a same amount of time. We appear to move in slow motion for them, while for tortoises vibrating at 15 hertz, we appear instead to move in fast forward, four times faster than them.

As for Seth in Jane Roberts, and the fact that there is no past nor future, but just a spacious present, where all of humanity's history happens simultaneously, would be the true definition of time, something that does not exist. But it still exists in the mind of everyone, even if the past and the future are constantly changing:

> "The past exists as a series of electromagnetic connections held in the physical brain and in the nonphysical mind. These electromagnetic connections can be changed. The future consists of a series of electromagnetic connections in the mind and brain also. In other words, the past and present are real to the same extent. No event is predestined, any given event can be changed not only before and during but after its occurrence. From within this framework you will see that physical time is as dreamlike as you once thought inner time was. You will discover your whole self, peeping inward and outward at the same 'time', and find that all time is one time, and all divisions, illusions."
>
> Jane Roberts, *The Early Sessions*

In this eternal now of Seth, if you were to visit the ruins of the most ancient city in Europe, the city of Knossos located on Crete in Greece, where the Minoan civilisation used to thrive, you would realise that they are actually living their history right now, at the very same time you are walking on their ruins.

It is also true, that several psychic mediums can actually switch to that other present at will, and witness it for themselves. It is often related to psychometry, retrocognition and time slips. It is like if psychics can pick up these electromagnetic signals, and re-create the past within their mind, a new bubble reality which co-exists with the present. Or, picking up that signal, that resonance of the interdimensional frequency of matter, from that particular time and place, can transport them there in the past instantly.

There are several astonishing examples, in the paranormal literature, of people who have suddenly found themselves in the past or the future for a while. Those events must also be explained by science. But if all the history of humanity, is more like a psychological construction in our collective and individual minds, or even a virtual reality, then anything is possible.

Especially if any event in our past, present or future, only exists as electromagnetic connections in an electrical universe, behind the scenes of our camouflage reality. It is pretty clear that Seth is describing what computers do, a virtual reality. Electromagnetic signals that our brain receives to create 3D bubble realities out there, instead of on a computer screen. All the patterns and signals are stored magnetically all around us, even in rocks, like memories stored on a hard drive. Which explains these ghost apparitions from the past, that reappear cyclically in certain locations, where these events took place. We can simply focus on them and fill the patterns with matter, to re-create any reality from any time period.

Hacking reality must be possible then, if one could access these electromagnetic connections or signals, and modify them with their thoughts. The biological computer here, is our own brain, although our consciousness seems to exist beyond the third dimensional or density world, which would explain déjà vu.

According to Dolores Cannon, in *The Convoluted Universe Book One,* we can see and experience the past, by simply vibrating slower than our normal vibrational rate. Just like vibrating faster, would bring us to the future. However, if we vibrate just a little bit faster, not enough to move to the future, time appears to run in slow motion around us. As we vibrate faster, time appears to stop on Earth. This has been reported by witnesses of UFOs and aliens. Vibrating a bit slower than normal should make time run faster, like from the point of view of tortoises looking at us.

I think the solution to this conundrum, is that the overall frequency of matter constantly increases in time, as everything expands. All particles are orbiting slightly faster as time goes by, and their orbit enlarges at a constant rate. This enlargement of their orbits is independent of any additional enlargement or contraction, due to a change in speed of the particles, because of more or less matter is being added or withdrawn from the system. It is not related to the crossover effect and the growth and shrinkage of the particles.

It is an enlargement of their orbits solely due to the constant inner expansion of the particles themselves, which accelerates all objects in the natural orbit effect, due to the relative motion of the geometry of expansion. I explained this in chapter 3, in section 3.11, subsection: *3.11c The third perspective - Expansion re-established after the relative effects - And how orbits enlarge in spirals and naturally accelerate objects.* But just like the expansion is unseen, because everything is proportionally expanding with everything else, this constant changing frequency of matter is also unseen by us, and not taken into account. We cannot notice or measure it, for us it remains the same.

In order to move to the past or the future, our vibrational rate needs to match the interdimensional frequency or resonance of the time and place we wish to visit. This is also true for instant travel within our own dimensional world, for example from London to New York in real time, since both these cities, at this very moment, are like two different dimensions or bubble realities within the same overall density world.

The spacious present is ever always changing, changing the past and the future. Our timeline, or perhaps timelines as Seth confirmed, are always in movement. Thoughts can affect them, as thoughts and emotions can change the vibrational rate of matter. Which is why when we hear extremely bad news, time appears to slow down. In fact, time actually does slow down in a way, because we vibrate more slowly, and we switch to slow motion.

It would explain why every vibrating life form is unique, since we all probably vibrate at a slightly different rate. And although we could always vibrate at that same rate, the overall frequency of matter is constantly increasing with the expansion of matter. This way we can also pinpoint someone or an event, vibrating at a certain frequency, and also pinpoint it in time, by locating it on the overall increasing frequency of matter in time.

What is important in the discussion above, with this ability they appear to have to stop time, by somehow making the particles in our body vibrate at a higher frequency, is not related to any real time relativistic effect, where the time rate would suddenly go faster or slower due to any acceleration in space or gravitational field, as described in Special and General Relativity. Anything related to the effect of time changing rates, must have an explanation unrelated to any real time relativistic effect as proposed by Einstein.

We also need to remember, that our entire physical reality is a psychological construction, exactly like in the dream world, and that all events are just electromagnetic or electrical signals of different frequencies. We are free to wake up in the morning, and imagine ourselves in any reality we truly deeply want, like one without wars or without an incoming Armageddon, or one where whatever happened in the past has no longer happened. Possibly by simply changing the vibrational rate of the matter composing us.

This said, we are all individually dragged down by the power of our beliefs. We need to convince ourselves, that the reality we need to create every day, is not the very one shared collectively by the whole of humanity, that we lived through the day before.

There are two types of time according to Seth, psychological time as in dreams, where time appears to run much faster than our normal time, and where we could potentially live an entire existence within an hour of dreaming. And there is our conventional time, all agreed by every single person in the world, with whom we constantly exchange information telepathically, just like several computers connected to a network. In this world, we have convinced ourselves that time exists, and we all live by the second, in a chronological and linear kind of timeline.

From Dolores Cannon's books, these other beings, including aliens or ultraterrestrials, can easily affect our mind to show us anything they want. They can change our entire surroundings, bring us to another planet, without even having to go there, and perhaps even they can bring us there instantly. Simply because they help us re-create that reality in our mind, just like in a dream, and it becomes as good as the real thing, as it is the same as the real thing.

Then, they are bringing us into psychological time, in a higher dimension, where you could easily be there for three days, and yet only an hour would have passed on Earth. As it is all but an illusion, like a dream, just like our physical reality is. However, it may just be the use of clever technology, speeding up the vibrational rate of our atoms and molecules, making them larger and more spaced out. And the use of lasers or higher density light beams, from a higher dimension, to create virtual realities or holographic images. Perhaps it is a mix of all of that.

It also seems we can travel anywhere in the universe instantly, in any dimension, in the past or the future, simply by knowing the interdimensional frequency of the matter in that location in time, by adjusting our vibrational rate or resonance to match. This is how alien technology appears to be working, from what we read in the books of Dolores Cannon.

It is not necessarily an illusion however. It does appear that an entire civilisation, living at a higher vibrational rate of matter, or the fourth or fifth world density, could experience a normal lifetime of 100 years, while a few days would pass for the whole of humanity.

Although, this is simply how differently time is being perceived, to the point that for them time does not really exist, in any concept or manner as it exists for us. There is still here what could be called a relativity of time, when it comes to comparing our time rates. This is called interdimensional time, in the books of Dolores Cannon, so I adopted the term.

Seth does say that the dream world, with its psychological time, which can run at any rate we wish, is closer to the reality of everyone else in the universe. As Dolores Cannon stated in *The Custodians:*

> "We are hampered by being trapped within our concept of linear time. It has been said we may be the only planet in the universe that has invented a way to measure something that does not exist. I have been told many times in my work that time is only an illusion, an invention of man. The aliens do not have this concept, and they have told me that man will never travel in space until he overcomes the erroneous idea of time. This is one of the main problems that keeps man trapped on the Earth."
>
> Dolores Cannon, *The Custodians*

Our idea of time is a collective invention, an agreed convention. If we were able to convince everyone, that we can create any reality we want in psychological time, a time that is not linear, we would have a very different physical world indeed, where no two days would ever be remotely the same.

No one then, could tell us what the future will be, and be so accurate, as to make it a self-fulfilling prophecy. There is not going to be an Armageddon or an end of the world. It is only the collective belief that it will happen, that will make it happen. So forget about it, despite all the prophecies in all the biblical books.

We are in charge of our future, through what we believe through our thoughts. But we are equally in charge of our past, and our prophecies can change, to become much more positive than what they are telling us now. As it is, these prophecies only reflect our present state of mind, they are only an indicator of where we are heading. But as surely as the future is not written, neither is the past. Only the present exists, and this is what they mean, when they invite us to live in the present.

And if we could, just through will alone, adjust our vibrational rate and find ourselves whenever we want, in the past or the future, at any given time and place, we would indeed have a totally different concept of what time is, as it would no longer be linear. Effects or consequences, could precede the cause of events, even, we could witness effects without causes, and it would all be normal. The past could be changed, the future could affect the past, and the spacious present would be all there is, as Seth states.

Freeing ourselves of our idea of time, is the first step toward opening our inner senses, according to Seth. We have our work cut out for us, learning to live more in psychological time, abandon the idea that our world must run on linear time. I'm not sure we could ever succeed, perhaps not on a massive scale at least.

And maybe, people in higher dimensions don't need to take time into consideration, because they have an equivalent concept to help them locate anyone, in any time frame. And that is, the rate at which matter vibrates, the resonance. If the frequency of matter is usually apparently constant, with the constant expansion of all matter, there is also an unseen constant increase of the vibrational rate of all matter. Consequently, you would not need the concept of time, if you were guiding yourself, and measuring instead, the changing vibrational rate of matter in time, in various locations.

Then, the constant and ever greater interdimensional frequency of the vibrational rate of matter, would be the equivalent to our concept of time. Except that none of it would be linear. You could go anywhere at any time, in the past or the future, and the past and the future would be constantly changing.

At least it would have a real basis in physics, since the vibrational rate of matter, depends on how fast smaller particles are orbiting larger ones. This in turn, affects the size of these particles and all objects. Adding energy, or more particles, to such systems, simply causes particles to become larger and orbit faster, spaces them out, and move them to a higher dimension or density world. While overall, they are constantly further expanding at a constant inner rate, in an unseen expansion, and also, in an unseen increase of the interdimensional frequency and enlargement of their orbits.

This is how we can move instantly to other dimensions, the past or future, to dreams, thoughts, our own imagination, and other locations where matter vibrates faster or slower. These are all other dimensions, realities made of matter and particles of different sizes, orbiting at different speeds, and creating worlds of different densities.

The interdimensional frequency of matter, simply means the number of orbiting cycles of neutrinos around and between the electrons, and of the electrons within and between electron clusters, atoms and molecules. The faster they orbit, the larger their orbits, causing a change in the size of the larger particles they form. This is just to highlight the difference with the other and more usual type of frequency of matter in our science, where the electrons, electron clusters, atoms and molecules don't change size.

The difference is more important when we are talking about the frequency of light or any electromagnetic phenomena, where the frequency may refer instead to the size of the electron clusters, meaning the number of electrons within the clusters, while the electrons have not changed size.

And for radio signals and microwaves, the difference in frequency only refers to a change in the vibrational rate of the compressed bands of electrons, within our own electromagnetic spectrum, where once again the electrons have not changed size, although they would still be orbiting faster in enlarged orbits, or slower in contracted orbits.

The size of the electrons is only affected when the neutrinos, within the electrons, orbit faster or slower, enlarging or contracting their orbits, causing the change in size of the electrons. This in both the second and third type of growth and shrinkage of matter.

9.4 John A. Keel, *The Mothman Prophecies,* Bigfoot, Mysterious Beings, Ghosts and Spirits, Dimensional Worlds and World Densities

We could say that I am a philosopher of theoretical physics, who is also interested in the paranormal. I've always been fascinated by *The Mothman Prophecies,* and recently I read several of John A. Keel's books. I realised then, that I had reached independently certain conclusions that Keel also reached in his books. It was kind of a confirmation of some of my theories.

I've been thinking about how we could explain ghosts and spirits, and for a while I kept thinking that they often appear like clouds of gases. I wrote an entire essay about how they could be made of gases, becoming solid when they materialise. Nevertheless, I could not explain how a gas could still keep the necessary molecular structure to support life, or some sort of consciousness. The very definition of a gas, is molecules in chaos. For a while, it seemed there could be hope with the fourth state of matter, plasma. Still, it looked highly unlikely, as it is still a gas.

Since then, after reading all the books of Dolores Cannon, and several of Jane Roberts, getting a hint here and there, but never any kind of readymade answer, I developed a better hypothesis. The difference between solids, liquids and gases, is how spaced out molecules are from each other. The faster the molecules are moving or vibrating, the more spaced out they become, and the more the material moves from solid to liquid to gas. This is just when we consider the spacing between the molecules, when the molecules always remain the same size, as they must do in today's physics.

Then I wondered what would happen if we could have different types of molecules of different sizes. What if what changed, when it comes to the paranormal - ghosts, spirits, UFOs, ultraterrestrials, higher beings - had something to do with a reality where the electrons were growing within the atoms, moving and vibrating faster, making atoms and molecules larger than usual, as well as more spaced out? Enough for light to pass straight through it all, like it is for gases, and suddenly we could no longer see or interact with that matter.

It would be like a gas, but truly you could still have solids keeping their molecular structure, capable of supporting life and consciousness. It would exist outside of our visible light, and even outside our field of interaction. In effect, I thought, this could very well explain worlds of different densities, of other dimensions. Other realities that exist right here on Earth with us, sharing the same space, but simply made of matter of a different vibrational rate, density and size of particles.

Turtles vibrate at 15 Hz, humans at 50 to 60 Hz, and flies at 250 Hz. It means all the particles composing these three different life forms, are moving at these frequencies, a specific speed. What are these frequencies corresponding to, I wondered? In my mind, it is how fast the neutrinos are orbiting within and between electrons, enlarging or reducing their orbits, affecting the size of the electrons. And in turn, it is how fast the electrons are orbiting within and between atoms, bonding atoms together to form molecules. The faster the neutrinos and electrons are orbiting, enlarging their orbits, the larger and more spaced out the atoms and molecules will be.

This interdimensional frequency is what defines the dimension or world density we live in. Every being at any specific time, and every location at any specific time, have a unique vibrational rate or resonance, which corresponds to the speed at which the neutrinos composing electrons, are orbiting and moving, changing the size of their orbits, which in turn defines the vibrational rate and size of the electrons, clusters, atoms and molecules, composing that specific dimension or density location.

This might be why the ultraterrestrials can move instantly anywhere in the universe, at any particular time, either in the past or the future. They can probably calculate what is the right frequency or resonance of a certain person or location, at any specific time. The past affects the future, but the past constantly changes, as well as the future, and the future can change the past. Everything is constantly changing, as everything expands and is in motion.

This is why they also say that time in a linear fashion, as we imagine it to be, does not really exist, if you can go to the past and change the present and future. We kind of live in bubble realities, more psychological than anything else, very much like many different interconnected dreams.

When you are standing in a specific location, you might think your frequency or vibrational rate never changes. However, matter expands at a constant rate, and in time, continuously expands a little bit faster than before. Hence, every second you are larger overall proportionally than before, and the matter composing you, and the location you are in, also constantly expand and rise in frequency.

The ultraterrestrials can probably pinpoint exactly where we are, at what particular time in history. They only have to adjust their vibrational rate to match ours, in order to move instantly there. Their particles then become of the right size and speed, or of the same resonance.

Reality is more like electric or electromagnetic signals or codes, that we all transmit and receive, and that the brain interprets, in order to recreate in our mind, the common reality that we see at any given time. You can easily influence the reality of someone, by suggesting anything through hypnosis, and create hallucinations or entire other realities. Because then, you simply emit those particular signals, that will cause someone to create these realities in their mind. At that point, you could say that it is as real as if it had actually happened, because it is probably the very same process that creates our actual reality.

When you move in the distance, you are not really moving, you are simply either shrinking or growing towards other objects or persons, that are initially all shrunk from your point of view. Everything is essentially located here at a same point. And things are only either enlarging or shrinking, from your point of view, as they approach or move away.

Very much like what someone would programme on a sophisticated computer simulation made of electrons, that would actually form into atoms and molecules. Everything is made of electrons in our reality, including protons, neutrons, atoms and molecules. In a virtual reality, they would programme a ship coming towards you as getting bigger, instead of moving in the distance, and once it has past you, it would simply get smaller. While for all that time, the ship was actually with you the whole time, simply shrunk or enlarged from your viewpoint.

The entire reality, every single location in the universe, any object or person, is ruled by the expansion of matter, which reduces the distance between everything. This is gravity, which also leads to the natural orbit effect. It is also defined by the growing and shrinking of electrons inside and outside the atoms, explaining electromagnetism, attracting and repelling forces, electricity, the density of matter and other dimensions. And finally, relative motion creates the illusion of distance, through the relative enlargement and contraction of matter, as objects approach or move away from each other. Overall, this is all due to the geometry of expansion, when we cannot see the inner expansion behind the scenes, because we are also expanding, as we are part of this universe.

A turtle moves in slow motion from our point of view, but from the perspective of the turtle, it is us who move in fast forward. It is the same for flies. By vibrating four times faster than us, they can achieve four times more than us within a same amount of time, plenty of time for them to see us coming with a fly swatter, and move out of the way. It could be said that flies are not existing within our human dimension. They are in a higher dimension compared to us, but a dimension that we can still see and interact with.

Any higher, and we get these mysterious spinning flying rods that we cannot see or interact with, but can sometimes be seen on photos or films in slow motion. They are most likely simply insects. Even higher, we get where the spirits and ghosts live, the world density 3.5. Or 2.5, if the mean spirit is from a lower density or dimension. And in the fourth dimensional world or world density, you have where the ultraterrestrials and their UFOs live, where they've always been living, sharing the world and the same space as us.

Now, this is where it becomes interesting. The second world density, where the matter is denser, and all the particles are orbiting and vibrating much slower, at very low frequencies, is the reality of the Bigfoot, goblins, gnomes and the Mothman. They are sharing the world with us, but we cannot see them and we cannot interact with them, because the spacing between their particles and their vibrational rate are different, and as a result, it is like a different reality, although still very much a solid reality from their point of view.

9.5 Known World Densities/Dimensions sharing the same space

1. Extremely dense and heavy life forms, black shadows?
2. Bigfoot, Goblins, Fairies, Gnomes, Mothman, etc.

2.5 Dark spirits appearing as dark black clouds, violent angry ghosts

2.75 Possibly our own physical body also exists here, which could explain how people could seem to create poltergeists without the need for any external influence. When we die, sometimes this physical body might continue on its own like an empty shell, or as a very confused consciousness, possibly moving to 2.5 and becoming an angry ghost

3. Our physical world

3.25 Our own physical body might also exist here, so when we die, this physical body continues on its own and possibly moves to 3.5

3.5 Spirits, ghosts of dead people

4. Ultraterrestrials, extraterrestrials
5. Other higher beings people appear to be channelling
6. Even higher beings? Archangels?
7. God?

The above table is just for guidance and ease of understanding, it could be slightly or fundamentally different. It is important to understand that every single person, every single location, every single particular time, is a dimension or world density of its own. There is in fact an infinite number of dimensions or world densities, but they can be grouped in the main bands above, when they are relatively similar in vibration and resonance.

For us, the fourth world density might appear like a gas, or a gaseous world, but for them it all appears very solid, since their particles are vibrating at that higher frequency, so they can experience that reality in a higher dimension. The word dimension here means the dimension of the matter composing them, and it is clear from the literature that, another dimension is synonymous with another density of matter.

John A. Keel got very close to this conclusion, when he talked about a different electromagnetic spectrum for the ultraterrestrials, where they live in realities of different interdimensional frequencies. The animalistic beings of the second world density, often have illuminated red eyes, because most likely, in that reality, they only see below our range of visible light, which is infrared light. While ultraterrestrials have huge black eyes like camera lenses, just like some insects, and the light they see must be above our visible light, meaning ultraviolet light.

I read somewhere that if you use ultraviolet filters on cameras, or somehow have technology that can see ultraviolet light, you could see a lot of UFOs out there in the sky, just standing there or moving about. And probably the same for infrared light, you would see all sorts of low intelligence creatures creeping around. Sometimes ordinary and FLIR cameras do get a glimpse of other dimensions.

However, it is not enough to simply look with infrared and ultraviolet light googles, filters or cameras. The ultraterrestrials' electromagnetic spectrum, as John Keel suggested, is not the same as ours. Remember, in their reality, all the electrons and the clusters are either smaller or larger, and are orbiting everywhere much slower or much faster than within our reality.

Heat and light are made up of clusters of electrons instead of photons, usually produced by heating an element. The size of these clusters of electrons defines the frequency of that light. The larger they are, the more they tend towards heat and infrared. The smaller they are, the more they tend towards ultraviolet, X-ray and gamma rays. The number of clusters passing within a second is the intensity of the light.

In our reality, or third density world, these clusters are relatively all the same size and equally spaced out, for one specific frequency of light. In the lower and higher dimensions, the density changes, the neutrinos change speed, and their orbits either enlarge or shrink, changing the size of the electrons. The electrons are then orbiting slower or faster within and between the clusters composing light, also changing the size of their orbits. This changes the size of the electron clusters and the spacing between them, and leads to different electromagnetic spectrums for these other dimensional worlds.

The electromagnetic spectrum should be separated in two different sections. The second section is the one I have already discussed, about heat, light and radiation, but the first one is something entirely different. While heat and light are clusters of electrons of different sizes, produced by heating an element, radio signals and microwaves are instead compressed bands of electrons produced through oscillating antennas.

If the emitted signal we are trying to receive, has been emitted using equipment from another dimension, then the signal would technically only be reliably received on equipment from that same dimension. Although sometimes, through flukes, we do receive these signals and communications, such as in the Electronic Voice Phenomena, also called EVP, where we can record, on any device, what appears to be disincarnated voices.

In order to communicate with beings of these other dimensions or realities, or in order to see and film them, we would technically need to change our very own vibrational rate, the speed at which all our neutrinos and electrons within us are orbiting and moving. The same for our equipment, be it radio, recorders and cameras.

When it comes to ourselves, it can be achieved through thoughts and intent, a willingness and desire to make it happen. When it comes to our equipment, electromagnetic fields (clouds of electrons) might help, air charged with static electricity might also do it, or a lot of heat or a lot of cold might help. Injecting a lot of electrons, through electron guns, might work. The more electrons there are in any location, the more excited all the particles will become. But there is no guarantee that any of it will actually affect the speed of the neutrinos, which is needed here to change the size of the electrons. Testing is required to find out what works best.

Perhaps the right tuning of radios might do it, without having to change the vibrational rate of the equipment. This is perhaps how these devices, we sometimes see ghost hunters using on TV, work. They receive signals made of clusters of electrons like light, clouds of electrons such as magnetic and electromagnetic fields, and compressed bands of electrons also called radio waves, with electrons of a different size, moving much faster or slower than normal. And when it comes to beings, we need to be able to hear and see them, while they are made of molecules and atoms of different sizes, vibrating faster or slower, and more or less spaced out than usual.

In areas where geomagnetic and magnetic fields are unusual, or near power plants and industries producing a lot of heat, or maybe in highly electrically charged areas, or whatever else, the barrier separating our dimension from the others may break down. This is why suddenly there are zones where we see a lot of activity, like the Bigfoot and the Mothman from a lower density, and ultraterrestrials and UFOs from the higher density, and probably also ghosts and spirits, which can be from either the lower or higher dimensional worlds, halfway through the other more concrete ones.

The lower the density they come from, the more animalistic they should appear to be. And the higher, the more intelligent, knowledgeable and powerful they should be. Before thinking of going to war with any of these species, it would be wise to recognise that we are most likely no match for them. They have proven many times to be able to disable our nuclear weapons at the press of a button. And chemical and bioweapons won't work on beings from another dimension. If they really wanted to annihilate us, we would not know what hit us, or how.

Most of them in the higher dimensions, according to Dolores Cannon, are benevolent, despite anything they might have done which might be questionable. This said, maybe they are not all caring and compassionate, even in the higher dimensions. This is much more complicated than what we could surmise, from reading all of Dolores Cannon's books. If you want the entire history, of all the other species of beings out there, and which dimension they come from, and the level of threat they may represent to us, you will need to read *The Wes Penre Papers - A Journey through the multiverse.* His *Levels of Learning* are free to read, and you would be hard press to find a better source of information concerning this subject.

Once something in the environment affects the rate at which neutrinos and electrons are moving within particles, suddenly we can all share, to a certain extent, the same reality or dimension. Which explains why sometimes there is a lot of activity in one particular location, what they call a gateway or an open window. And I'm sure some of these beings from other dimensions, can get stuck here. While the great percentage of people who go missing every year in the US alone, might very well have moved permanently into these other worlds. Then, they are not dead, and I do wonder how they adapt over there.

The true reality is that in the same space, we share this world with another reality made of a lower density of matter, and also another one of a higher density of matter. Where molecules are either smaller or larger than normal, and less or more spaced out, with particles moving slower or faster.

In these worlds, there are probably entire populations of these beings, with cities and UFOs and ultraterrestrials everywhere. And it is only when the barrier between these realities disappears, through most likely natural phenomena, or through the action of these ultraterrestrials, who appear to be capable of lowering their vibrational rate at will, but never for very long, before they must return to their world, only then can we share the same reality.

Anything that can affect the speed of the neutrinos orbiting the electrons, inside and outside the atoms, will do it. By injecting more neutrinos somehow, within and around electrons, even within the atoms. It seems like high electromagnetic fields can do this, but it might also be dangerous.

It is likely that all human beings are made of at least two superimposed layers of matter. One layer of matter moving at normal speed in third density world, which is our physical body, and another made of matter vibrating at a higher rate, of the density world 3.5, which is also a physical body. Which means that we are multidimensional, and kind of live in at least two different dimensions at once, without being fully aware of it. The dream world is probably in world density 3.5. It also appears that both our physical bodies share a consciousness, and are functioning and capable of storing memories.

Our physical body might also be made of a third layer of matter, vibrating at a much lower rate than normal, closer to the world of the Mothman. When we are happy, we move towards the higher dimension, and when depressed, towards the lower, as emotions affect our vibrational rate, and also our time rate.

When we die, quite possibly one of these physical bodies can continue to exist on its own, sometimes normally, other times perhaps in a diminished manner, if the death was violent or traumatic. People don't change after death, if they were angry violent people, their second physical body that survives, is likely to be the same. If they start vibrating more slowly, they might move towards the denser reality, to the world density 2.5, if faster, towards the higher dimension 3.5. But it does appear that sometimes, they do change after death, and become more compassionate and understanding than while they were alive. These people were probably not that bad to begin with, but they probably hid it very well.

Rituals and psychological preparation before attempting communication with other dimensional beings, probably has or should have something to do with raising our vibrational rate, so we can attract and communicate with beings in the higher dimensions instead of the lower ones.

Keep in mind that, the day we successfully develop the technology to communicate with all these other beings, ghosts and spirits, or find a way to see them, is the day we will effectively raise the dead back to life, in the biblical sense. They were never truly dead to begin with, and there must be a way to bring that second physical body, living in higher dimension, back to our dimension. We only have to reduce the vibrational rate of their matter, or raise ours. And this is what psychic mediums probably do.

Now we can understand why the ultraterrestrials are playing these mind games with us, leading us astray constantly, lying and deceiving us. They are simply protecting themselves, as we must be a threat to them. They give us hints of our future that they can actually see, since they are vibrating faster than us, and relatively speaking they can achieve much more than us in a same amount of time.

Essentially, their time is running relatively faster than ours. Not only can they see our future, but they can lower their vibrational rate to specific times and places, in order to tell us our future. John A. Keel related several of these stories, including in *The Mothman Prophecies.* He said that eventually they let us down, and the future they tell us is incorrect. It destroys our credibility, so nothing we say will be listened to by others, and they render the entire idea of what happened completely ridiculous. The absurd, when it comes to any encounter with them, is common. They may simply be trying to hide the truth from us, so they can live in peace.

The time rate difference also explains why some people seeing UFOs, and interacting with ultraterrestrials from higher dimensions, suddenly see everything around them in slow motion, or even disappear from view to everyone else around. They must be in a bubble where all the particles are moving much faster, effectively existing in a higher dimension, leaving our dimension in a kind of frozen state that they can still see. Several such accounts were related in Dolores Cannon's books.

These ultraterrestrials might live in complete fear we will find a way to break the barrier between the realities, find a way to interfere, and perhaps even go to war with them. Not surprisingly, every nuclear bomb we detonate must be as devastating to them in their reality as it is to us. No wonder we caught their interest when we started testing these weapons. These kinds of explosions must destroy matter in all dimensions, I reckon, not just ours.

In a higher dimension, their reality might be more fluid, ultralight, with less gravity, with animals and plant life unusually large compared with ours. Still, it is all made of matter, solid from their point of view, and it shares our reality the very same way the air and gases in our atmosphere do. And for the Bigfoot and the Mothman, they also share our reality the very same way water, rocks and dirt do.

By misleading us, making us believe they are from other solar systems and galaxies, from other planets, and even from particular constellations (while this makes no sense, since these stars composing constellations could be millions of light years apart from each other), they are hiding their true location, they are protecting the truth. Which is that they live side by side with us, intertwined and interconnected to, and within, the fabric of our reality. And that it would not take much for us to develop the technology required to change all three interlinked realities forever, including the lower one of the Bigfoots and the Mothpeople.

Now that we have the correct understanding of the physics of our world, as I believe we are getting closer to, it might become easy, within years, to find ourselves in these other dimensions. One more revolution in theoretical physics and we're there, and perhaps this present book is just it.

This said, possibly not all ultraterrestrials are from Earth. Certainly, other planets of our solar system could sustain life in other dimensions. It has been said that the gas giants Jupiter and Saturn, at the far end of the solar system, are actually four-dimensional worlds, and are inhabited. Instead of gases, they may very well be made of molecular structures of different sizes of molecules. They may appear gaseous to us, but possibly they are not. However, for beings living there, in other dimensions, they would see these planets as solid as we see ours. They might be paradises. They may also visit us here on Earth, who knows.

And God only knows about aliens from outside the solar system, they've lied so much, it is hard to distinguish the truth. Have you ever wondered, why any knowledge coming from a great channelled source, never quite gave us any answer, that could advance our knowledge and discoveries by leaps and bounds? Are we that stupid, is our reality or science that different, or have they only ever provided half answers, hints? I can understand if they feel we must discover this on our own, and hopefully we will, but then overnight our existence will most probably radically change.

I've worked most of my life on a new theory of everything, which should explain everything, including the paranormal. And it starts with what is gravity. Put very simply, gravity is due to the fact that all matter constantly expands while we are unaware of it, because we proportionally also expand along with everything else.

This expansion of matter is the reason why the distance between objects is always reducing, what Newton stated was an attractive force acting at a distance, when really there is no such force acting at a distance. We are not attracted to the Earth, the Earth constantly expands underneath our feet, pushing us upward. A falling ball from a plane is not falling, it is free floating in space until the Earth, in its expansion, catches up with it. Although everything expands at the same constant rate, the distance reduction between objects actually accelerates. This is what gravity is.

It may seem crazy, to think that atoms are constantly expanding at the same universal expansion rate, no matter the dimensional world we are in, and this expansion reduces the distance between objects and explains gravity, but hopefully Mark McCutcheon's book *The Final Theory,* and this present book *New Age Physics,* will prove convincing. It makes sense, and it resolves every single big puzzle our science has been dealing with since forever.

The only reason the UFOs can fly so easily, defying our laws of gravity, is because they are made of matter vibrating faster, hence of matter of a lesser density, possibly even less dense than the air or gases in our atmosphere. It is easy to fly out of the atmosphere when you are made of such matter, because what is less dense is always automatically pushed above what is denser. Just like a balloon in water being pushed to the surface, or a balloon full of helium climbing in the atmosphere. Helium is the least dense gas in the periodic table of elements, after hydrogen.

The only way for us to achieve anti-gravity, is to build ships vibrating faster, to the point that all the particles will move to a higher dimension. But not necessarily one where the ships would disappear from view, and that we could no longer interact with it. Maybe the dimension of insects would suffice. They can easily fly, since they may be made of matter less dense than air, through larger and more excited molecules and atoms, with neutrinos and electrons moving faster, with enlarged orbits.

And the mystery of the flying spiders covering miles on the oceans and reaching boats at sea is finally resolved, despite the fact that they don't have wings. Or how bumblebees can fly with their very small wings, seeming to violate the laws of physics. Ever wondered how insects can walk on the ceiling? They are made of matter of a lesser density, and the ceiling is constantly expanding downward every second, which is enough gravity to keep them there.

9.6 Is humanity truly being enslaved by beings from other dimensional worlds?

When we read *The Wes Penre Papers - A Journey through the multiverse,* we realise that we have been involved for a very long time, one way or another, with a myriad of species of beings from lower and higher dimensions. In fact, there's never been a time when we have not been involved with them. Our entire mythology, religious bibles, literature, and even children's fairy tales, are filled with our encounters with these beings.

We have called them by several different names over the millennia, and euphemisms such as aliens, monsters, gods, angels, fairies, goblins, demons, the devil, Satan, evil, etc. And yet, it is poorly understood who these beings truly are, and what is our relationship with them. For most people, they don't even exist, it is all fairy tales, delusions and conspiracy theories. But what if it is not?

I am not talking about a possible fake alien invasion, that allegedly Wernher von Braun, the German-American aerospace engineer, warned us about on his deathbed. I'm talking about a possible enslavement of humanity, by one or more of these species from other dimensional worlds, and what they are planning for us. If this is truly the case, as we hear from many different corners, we all need to be made aware of it, so we can do something about it. But how true is it?

I had no wish to become political in this book, but if it involves beings from other density worlds, trying to take over humanity, or having already done so, we are no longer talking about politics. Unless we call this interdimensional politics, or interdimensional foreign affairs, which then becomes relevant to this chapter. And of course, it is not all races of ultraterrestrials who work against us, some are actually benevolent and are helping us.

So far, we have proven unable to stop our leaders from doing just about anything they wish, and these days our leaders are rarely wise, intelligent or reasonable people. At the turn of 2020, we have witnessed how alarmists and extremists they can become, and easily manipulated into creating self-made crises leading to catastrophe.

Our leaders destroyed our economy, the supply chain, our food and energy sources, our health and health services, our standard of living and our way of life, all without a second thought about what they were doing, and without even any studies or impact analyses, to foresee the consequences of such global actions.

As a result of the outcry, the right to protest and assemble is being suppressed. Freedom of expression has entirely but disappeared, everyone is being censored and deleted. I fully expect that fairly soon, it will hit theoretical physics and every other field of knowledge. I have already seen harmless people banned, for talking about the Flat Earth Theory. This is where it starts, with theories most people agree are incorrect, so no one cares, until they gradually move on to ban everything else.

Schools, shops, restaurants, pubs, concerts, libraries, cinemas, theatres, museums, all gone, and most are bankrupted, so gone for good. Families unable to pay the rent are now homeless, others have to sell their properties, because of loss of revenue from rent. Millions are starving in what used to be the rich countries. Wearing masks everywhere is now mandatory, even although there is over a hundred studies showing how useless they are in a pandemic, and many more studies proving they lead to serious health issues.

And now we need permission to travel. We must navigate a nightmare of a system, with quarantines and health passports, photographic evidence of where we are daily, and endless testing every other day, costing a fortune. Hospitals have been shut down, denying many of urgent treatments, and condemning them to death. People in care homes were left isolated, denied medical treatments, and even put on end-of-life medications. We have also, more or less, all been prisoners within our own homes, with not even the freedom of movement, and coerced into dubious medical experiments. All of it based on flimsy evidence and shady studies, from corrupted and untrustworthy people and organisations, making billions if not trillions in profits. Can you really put it any other way? Or still defend that it was required, somehow?

What are the definitions of totalitarianism, fascism, oppression or tyranny, no matter the excuse that could justify such a state of affairs, if this is not it? Have the definitions of these words changed recently? We cannot tell what the future holds, but it is certain that the world will never be the same again. And as it is only getting worse, civil wars and serious crackdowns are probably lurking just over the horizon.

It is as if, all our leaders were either bought or threatened in some way, or both. They appear to have no freedom to make any decision benefitting the people they are supposed to represent. And now, all the leaders worldwide seem to talk with one voice, they all say the very same unconscionable things, no matter their political inclinations.

The very same speeches, laws, rules and policies, are passed in every country in the world, under emergency measures and without oversight, all at the same time in a co-ordinated effort. So-called experts, everywhere in the world, all agree on everything, and if they don't, they are ostracised, discredited and silenced. All news channels and newspapers worldwide, in all languages, are all parroting the exact same propaganda, over and over again.

It is pretty clear, that there is now one world government hiding somewhere, which also controls the worldwide governments, mass media, the military and the police of every country. We can only hope that soldiers and police officers will wake up, and work with us, instead of against us.

What a powerful propaganda machine, led by a team of experts in the psychology of behavioural change. It works perfectly well on the majority of the population, there is no more doubt about that. This is not new either, we witnessed these kinds of things many times before, even within my lifetime. It is just that today, it has become painfully obvious and undeniable. These leaders are not speaking their mind, they are following a script written by others from one source, without being allowed to deviate. We must be nearing where this is all leading, since they are no longer hiding their agenda.

On the surface, central and world banks, large corporations, agencies of the UN, NGOs and other think tanks, appear to be in charge of the world, in a technocracy hiding behind fake elections that no longer matter. That much I can see for myself. But there is a lot of information out there, stating that there could be a smaller group of extremely rich beings, in control of all these organisations, who are not even from our dimensional world. They would come from either the lower or higher dimensions, and ultimately their agenda is one of genocide, preventing births through sterilisation and abortions, and eugenics, but always presented under the guise of public health, public protection and saving lives.

All new policies, rules and laws, usually achieve the opposite of what they are meant to achieve, and always seem to lead to less rights, liberties and freedom, and ultimately to more deaths. To the point, nowadays, that if the politicians or the mainstream media say something, you can bet that the opposite is true. Everything is reversed, but the conditioned majority remains blind to it. It is all very simple and clever, deceptively so.

If some beings from other dimensions, see humans as easily manipulated through basic psychology, and as useful animals that can be exploited, if they only keep the best specimens and their progenies, then it is understandable that they have no compassion or humanity for the human race.

Considering how we treat our animals, we are not that different. Would you not want to ensure that all your cattle and sheep are tagged and microchipped, so you can bring them back to the flock when they go astray? And that they have no rights or liberties whatsoever, so you can do whatever you want with them? But you might allow them to visit other fields, every once in a while, if they get a special permission, official form of identification, photographic proof of where they are at all times. While making it very complicated, as to discourage such behaviour. Because, it is far easier for you, if they remain at all times in their allocated small space, and not be allowed to say anything or complain against you to anyone.

And of course, only the best cows and sheep should be allowed to reproduce, so we can obtain the best pedigree. While using drugs and injections to cull the older ones, the sick and the weak, the ones without much of an immune system, which are easily killed. Better sterilise the dumb and the bent ones. Why would you keep costly dead weight and useless eaters, within your flock of cows and sheep? There would be no point to it, if they cannot fulfil their purpose, the why you keep them in the first place.

However, why kill only a few million people, when there are eight billion of us? Unless eventually, given the chance, they will ramp up their effort, and kill worldwide all the old, sick, mentally ill, gay, stupid, handicapped and lesser races, in their opinion. Ultimately, to save their precious resources, and the planet from a change in the weather, they might be planning to kill billions of people.

I don't believe for a second that the change in the weather is manmade, as it is simply normal cyclical variations in weather patterns that come and go. We have data from the last hundred years which shows similar patterns, and no one was alarmed then.

These beings don't strike me as careless caretakers, who would randomly kill people in the hope that the good ones survive. They intend to keep only the best of us, which is why they need the genome of the entire population of the world. Through these processes of injections and health passports, and especially with their PCR tests, and without our knowledge or consent, they are building the global genomic database, which will decide who has the best genes, and consequently, who lives and who dies.

Some people might be extremely intelligent, talented and skilful, and went on to change the world, but their genes might have shown nothing remarkable, and even, would have disqualified them from living. Who knows what kind of genes we need to have, to be useful for their purpose?

Perhaps genomics is all that it is cracked up to be, and maybe it will save many lives, and prevent many more from seeing the light of day, since they will know everything about the babies before they are even born. But it is the other potential uses of genomics which are worrying, and the fact that it has been collected in a dishonest way, it suggests bad intent.

As to why they need us, it has been said some species appear to attach themselves to us and feed on our energy. They apparently need negative or low vibrational energy that human fear, hatred, anxiety, anger, and depression generate. And we seem to be providing other species with natural resources, like mining for materials they require. And what else do we do for them?

You will notice that their plans are very long term. They have been implemented decades ago, often even in previous centuries, and this is why history constantly repeats itself. As it is always the same plan, being implemented decades after decades, over the centuries, by the very same beings, who can probably live for a thousand of our years, especially if they are from the lower dimensions, or operate from there. And we are powerless to stop history from repeating itself, every time that it happens, even when we are aware of it, because it is all extremely well planned. They have the benefit of knowing why it failed before, and can anticipate these variables every time they try again.

And as long as it ends up killing a lot of people, preventing many undesirable births, and bringing more control and coercive measures, then it is a success. And it does not matter to them, if every single person who was bribed or threatened to make it happen, or any person who appear to be running their show, ends up under a bus, in prison, or at Nuremberg type trial. Every single one of those responsible, no matter how rich, famous or important they are, are expandable to them, and a potential patsy ready to take responsibility for their own crimes.

It is quite a game of chess, when your funds are unlimited, and when you can plan to lose all your pieces, and still win. So, none of their plans ever fail, it is just a question of how long they can get away with it, before we find a way to stop it. Hence in time, it becomes more extreme, and we find out afterwards that there were millions of deaths as a result. Just as it happened many times before in recent history. And the day they finally succeed, and that there's nothing left we can do to prevent the takeover, the greatest genocide ever will probably follow.

Unless it is essential, it is probably best to avoid medical doctors and hospitals, and especially their drugs and injections. In this day and age, and for quite a while now, it is all about genocide. Prescription drugs is officially the third leading cause of death in North America and Europe, but unofficially it probably is the first. It would not surprise me, if dementia and Alzheimer, were mostly a consequence of the drugs they provide, and that, implicit everywhere, are the 'do not resuscitate orders' and euthanasia.

Hopefully a new breed of doctors and nurses will come about that we can trust. It usually takes years to organise and fight that sort of thing, plenty of time for the policies to become severe and create total chaos, which usually will take decades to fix and reverse. By which time, they're ready for the next global crisis to be unleashed.

These beings from other dimensions, seem to know exactly what is going to happen next, through great predictions and training exercises, that turn out to describe events identical to what actually happens soon after. It is as if they could see the future. It could simply mean, that they are often successful at implementing their plans, or, they literally know the future, having mastered going to even higher dimensions, where the time rates are relatively faster than ours. They are in no hurry, remember, time is meaningless to them. In the lower dimensions, they move in slow motion compared to us. Decades for us could be years or months for them. They are like tortoises or reptiles frozen in time, from our point of view.

It is high time that we figure out and master, the real mechanics and physics of this universe. Or we will continue to be ignorant and easily exploited, by beings of other dimensions, often called evil and demons in religious circles. They have a full understanding of how reality really works, and have no ethical or conscience issues, while taking advantage of it against us. Because for them, we are just useful animals that need to be brought under their full control. Preventing us from thinking at all, and expressing our opinions, seem to have always been one of their goals. Along with the one to know where we are, anywhere in the world, and with whom, at all times.

Why are they using such intricate subterfuges to deceive us into submission? Why try to control the masses, through the illusion of democracy and freedom, when it is simpler to control them with the threat of annihilation? Because, with an open threat, we would rebel, we would develop a resistance, we would refuse to do anything they need us for. It would be war, and we might win against them, or at the very least, become useless to them.

What better way to establish your supremacy, when no one knows you even exist, and that you can never become a target people could organise against? While people feel they are free, living in a democracy, going on with their lives, believing that nothing is wrong? A lot of thought and organisation went into this, for a very long time for us. But past a certain point, it comes a time when all the illusions vanish, and it becomes perfectly clear in what kind of world we live in. One without any democracy, rights or freedom. We are always truly ruled through fear and psychological manipulation.

It is possible that they are more powerful than us, and could annihilate us easily in our sleep, but can't do it because they need us. Either way, fighting them is a big risk. Bringing public awareness to this state of affair, and negotiation, might be the solution.

Interdimensional politics, if they are willing. But why would they be willing?

You don't negotiate with, or kill your cows and sheep when they rebel, you just find better ways and more radical means, to bring them under your full control. Especially if they are no threat to yourself. First, most people don't know interdimensional beings exist, they don't even know if the idea of other dimensional worlds is possible. And finally, these beings don't even have a real tangible materiality in our reality, we can't even interact with them or see them. We are indeed no threat to them at this stage of our evolution, at our level of knowledge and development. Hopefully we can change all that.

But they made sure we remain ignorant, about the true nature of our reality. Anyone telling the truth, is dismissed as a pseudo-scientist, a conspiracy theorist, or a nutcase ready for the asylum. Anyone presenting a theory in physics that goes against Standard Theory, is always being ostracised, and cannot expect peer reviewed publications. Not many working on theoretical physics are free to speak their mind. They certainly could not say anything against Einstein, or utter the words UFOs, aliens or demons, as they would be ridiculed out of the building, lose their job, and any funding they might have. I nearly avoided the topic myself, but self-censorship is the worst form of censorship.

If we are not allowed to speak about these things, how are we meant to fight them, advance and move on? We need to get to the bottom of this. There is already enough proof out there, which need to be brought to the attention of the public, while attempting to get through their conditioning against such ideas, before it is too late.

Then again, perhaps it is all fairy tales, delusions and conspiracy theories. I guess it is up to you to investigate the subject further, in order to make your mind up. Reading John A. Keel, Wes Penre and David Icke, seems to be a good place to start. And remember, you first need to get through your conditioning against such ideas.

9.7 Conclusion - Five types of expansion and contraction of matter to explain the entire physics

As stated at the end of the long summary in chapter 2, the five different types of expansion and contraction of matter explain the entire physics governing all realms of existence. I hope I have presented a convincing picture of a complete theory of everything in physics, entirely based on the expansion and contraction of matter. This could be what the channelled material in the New Age and spiritual literature has been referring to all this time, and finally we can understand what they meant by everything is expansion.

The fourth type of expansion, is the psychophysical reality we create out of our mind, all around us every time we wake up in the morning or move to another room. Once all the matter is out there, in the first type of expansion, neutrinos, electrons, atoms and molecules, expand at a constant universal rate, raising steadily the frequency at which all these particles orbit and vibrate in the process, while all orbits behind the scenes are constantly expanding in enlarging spirals.

The fifth type of expansion and contraction of matter is relative motion. Objects are not covering any distance, they simply expand as they move towards you, or reduce in size as they move away from you. Like it would be in a psychological reality, a computer simulation or a holographic universe.

This is from our own point of view only, as it is all relative motion. Which means that the particles within the matter are not, as a consequence, truly expanding and reducing in size, this is just how we see it. There are no mechanisms in this process, where the particles suddenly truly contract or expand, and where internally the spacing between the particles would somehow change, unlike for the other types of expansion and contraction of matter.

The second and third types of growth and shrinkage of matter, are about the orbits of neutrinos around and between electrons, and the orbits of planets around and between solar systems, linking them together. It is also about the orbits of electrons around and between atoms, and the orbits of solar systems around and in between galaxies, also linking them together.

The expansion and contraction in size, of these orbits of smaller particles or objects, linking together other larger particles or objects, is creating an additional attracting and repelling force between these objects. A force capable of counteracting the gravity, that is due to the inner expansion of matter, keeping all these particles and objects apart.

This leads to the crossover effect Mark McCutcheon explains in his book *The Final Theory*, the second type of growth and shrinkage of matter. For McCutcheon, the electrons freed from the atoms, can suddenly expand a great deal compared with the ones still in the atom.

For me, it is just that the orbits of the smaller particles, in this case the neutrinos, within and between the electrons, are suddenly enlarging a great deal, enough to push the electrons out of the atom. But never expanding that much, while they are still interconnected with other particles within the subatomic world, as the entire system, through the geometry of expansion, balances all the varying orbits.

In electrical fields in static electricity, the orbits of the neutrinos can stop enlarging or reducing altogether, forming walls or clouds of electrons, still constantly expanding at their normal expansion rate against each other, however appearing immobile from our point of view. They can remain like this forever, if left undisturbed. Like the repelling force keeping two 'charged' glass rods apart, before an electric discharge destroys the field of electrons between the rods.

And when the electrons are completely freed from the subatomic and atomic realm, and start expanding freely in space, either as clouds or clusters of electrons, then the orbits of the neutrinos must be as large as they can be, before the orbits lose cohesion. And in time, as magnetic fields, light and heat travel in space, eventually these orbits lose cohesion and the particles disperse.

In the second type of expansion and contraction of matter, the electrons are growing to leave the atom, to become all the different types of energy identified in science, or shrinking back to the atomic realm, as the neutrinos slow down and their orbits shrink, reducing the size of the electrons.

The third type of growth and shrinkage of matter, is related to the second. It is also the acceleration of the neutrinos and their orbit enlarging, however, without actually leaving the atom, and without going through the crossover effect. When electrons within the atoms are growing or shrinking, the atoms become larger or smaller than normal, creating larger or smaller molecules than the ones we are familiar with, and changing the vibrational rate of matter. This creates realities of different densities, other dimensional worlds that we cannot see or interact with.

There is no guarantee that other planets have solids, liquids and gases made of molecules and atoms that are the same size, vibrating at the same frequency of matter, and similarly spaced out as on Earth. Although the planets closer to the Sun, are probably of a similar density, both in terms of normal density, which is the quantity of mass per unit volume, and also interdimensional density, which is when the electrons, the atoms and the molecules are larger or smaller than normal, vibrating at varying rates, and more or less spaced out. It is possible that the larger gas giant planets, further down in the solar system, will uncover surprises.

The third type of growth and shrinkage of matter, can also happen outside the atoms. If electron clusters, electron clouds and electricity, are produced from higher or lower dimensional matter, or from matter of a different interdimensional density, then the resulting energy would also be different than our usual one, and we might not be able to see it, measure it or interact with it. Or if we can, it could be lethal.

This interdimensional energy, would be made of electrons already larger than normal, while within larger atoms of a different interdimensional density of matter, creating clusters and clouds of electrons in other dimensions or world densities. The energy in the fourth dimension or fourth world density, is made of clusters of electrons that are made of electrons orbiting faster and larger than our normal ones, and these clusters and clouds are larger, with a higher vibrational rate, and more spaced out as a result.

Somehow, the neutrinos and electrons within the clusters and the clouds, must be orbiting faster, their orbits must be larger, and everything overall must be slightly larger than normal. This interdimensional energy or light should not truly affect our third dimensional world, but possibly all of this constantly intermixes. Each dimensional world must have its own electromagnetic spectrum, at different scales than our normal one.

These five types of expansion and contraction of matter are all interlinked together, while being distinct from each other. If you try to visualise these five types of expansion in action, all ongoing at once, you will find an extraordinary and interesting way of seeing the bubble universe that we each create around us.

Whenever electron clusters are freed from the subatomic world on the Sun, and heat and light are launched into space and take eight minutes to reach the Earth, or whenever a spaceship is launched from somewhere towards Earth, both the clusters and the ship expand in several different ways. They expand in all five different types of expansion and contraction of matter all at once, they never really move or cover a distance.

Other stars are extremely small in the tapestry of our reality, the electron clusters must expand for years, for us to see them from that scale. They are not covering a distance, they are expanding towards us. And our acceleration or deceleration, away or towards that light, or away or towards a travelling spaceship, will relatively change the rate at which they expand or shrink compared to us, compared to our frame of reference.

Motion is relative, and in effect there is no more motion or speed in this reality, since the universe has no distance, no space, no time and no spacetime. Ultimately, everything is here with us, everything we can see in the universe, is just shrunk from our point of view. This is basically already applicable to a virtual reality built by a computer. Which is possible, if as Seth states in Jane Roberts' books, this entire physical reality comes out of our brain, and is constantly being rebuilt with the energy (electrons) permeating the electrical universe at the base of our physical reality.

So far, everything in the universe is explained, by the expansion of matter from the centre of our brain, to create our reality, and the expansion of matter to justify gravity, chemical bonds, nuclear forces, electromagnetism, orbits and all energy phenomena.

Energy is just growing and shrinking electrons, pushing against each other on wires for electricity, in clouds for magnetic fields, in clusters for heat and light, and in compressed bands for electromagnetic fields, such as radio signals and microwaves.

And now, any object moving into this reality, is just further expanding or shrinking from our point of view. Everything is explained by the expansion and the contraction of matter.

Mark McCutcheon does not agree with me on all of this, but he certainly confirms that motion is relative, although, as I said, more like Galileo's relativity. As for all objects made of atoms in the universe, they never stop expanding proportionally to everything else, they still double in size every 19 minutes, as calculated by the Atomic and Subatomic Expansion Equation derived in Mark McCutcheon's book. The value of this universal atomic and subatomic expansion rate, for all neutrinos, electrons, atoms, molecules and objects made of them, is 0.00000077 $/s^2$. Note that there is no metre in this value, it is a proportional expansion rate of matter in time.

We each are the centre of our world, the return to an egocentric view of the universe. The main basic new principle, is that everything in physics is driven by the expansion of electrons and atoms. Now, can we really dismiss a fully working theory of everything, that even explains and take into account consciousness, as our very mind builds the physical universe we exist and evolve in?

Everything has a consciousness, according to Seth, electrons, atoms, molecules and cells, albeit a limited consciousness. But also moons, planets, stars and galaxies. Matter is conscious, so is the energy permeating all universes, since it is made of electrons. And consciousness also has a memory, so matter is conscious, can store data, can remember and decide, and be told by our consciousness, how to shape and form, and if to remain as such or change.

In his book *The Holographic Universe*, Michael Talbot talks about how each one of our organs has a consciousness, and are affected emotionally depending on how we treat them and talk to them. If we can communicate and control every single particle within us, and any groupings of such particles, which lead to incrementing larger consciousnesses, responsible for other smaller consciousnesses, then we could ensure we never get sick and live forever. Hypnosis is a way to achieve this communication and healing, but I see no reason why it should not be a conscious process.

Under hypnosis, I once spoke with a sun. The person I hypnotised, was living several parallel existences to this one on Earth, including being a star somewhere in the universe. It caught me so unprepared, I did not even ask where in the universe it was located. In the end, I said, are you not bored with this existence, if all you do is stand there in space, moving about? The answer came immediately. If the person under hypnosis was making that up, no thinking was required to obtain the answer. The sun said something to the effect that, it could feel and experience the existence of everything within its system, and live through every single consciousness composing it, and it seemed like a fulfilling and meaningful existence. More than mine, that's for sure.

I can't even explain to myself, how I could have thought for one second, that I was leading a more interesting and less boring life than the one of a sun. There is so much to explore within each of us, that being in prison is no indication of our freedom to live a meaningful existence. I just wish it would be easier to access all this information about ourselves. But if it were, we could easily forgo to actually live and acquire new experiences in this lifetime.

Determinism and fatalism are unlikely, in a universe where everything is conscious and capable of changing the behaviour of particles and objects. Nothing is pre-determined in this universe. Particles and astronomical bodies, are not simply following their pre-determined paths, that we could calculate accurately over millions of years, even if sometimes it might seem that way.

This conundrum nearly drove me insane, when I was a kid. I thought, as soon as particles are set on their course, then they must follow their pre-determined paths, which might or might not lead to collisions and destruction, and nothing could change that. And so there could not be free will in this universe. But thoughts, free will, consciousness, can change anything at any time.

And although, so far, we are lucky that nothing much has changed in our night sky, it does not mean that one day it might not radically change. And suddenly, the Earth's orbit around the Sun, could significantly enlarge or reduce, along with the orbit of everything else in our solar system and the ones nearby.

The destruction of one planet, in any of the solar systems interconnected with ours, could significantly change the orbit of the Earth and lead to our extinction. If we don't start populating other moons, planets and solar systems now, *en masse,* we probably won't survive.

The solution is not to reduce the population of the planet, as I am sure Earth could easily accommodate twenty times our actual population, without much trouble or lack of resources. The solution, is to get out there in the universe in large numbers. And I hope this new physics is just what we need to jumpstart this exodus.

Our mind is supposedly outside our brain, outside our physical body. Although, an extremely small mental enclosure, at the centre of our brain, is what builds the physical reality, from what our mind or consciousness tells the brain via electrical impulses, from what Seth calls mental genes and mental enzymes.

Memory is not exactly in the physical brain, or within matter like electrons, atoms, molecules and cells. Memory would be in their consciousness, the part of their mind which exists in the electrical universe, meaning in the superimposed layer of energy made of neutrinos and electrons. It would have to be, if our entire physical reality is just camouflage, as Seth states. In the end, reality is all just an interpretation and a construction of our brain.

The centre of our brain is certainly an important place for us, for our physical reality, according to Seth. And he says that, although reality is composed of real locations, just like the locations in our dreams are equally real, or any locations we can imagine in our mind, they occupy no real space, because it is all psychological and mental in nature. He also says, that mental genes and mental enzymes build reality. So it is very much a biochemical and bioelectrical process, furthering the idea that it all happens in the brain, although this brain is part of the camouflage reality. It seems that nothing about us is from this world or dimension, really.

The advantage of Atomic and Subatomic Expansion Theory, is that it describes the physics not only of our physical reality, at whatever scale or world density we choose, but also of all possible realities and planes or systems of existence, including the dream world and the world of our imagination. It all works on the very same principle of the expansion of the neutrinos, and every other particle as a consequence, and the growth and shrinkage of electrons, in realities made entirely of neutrinos and electrons. The neutrinos, like the planets, are the only particles that are neutral, because they are not made of a nucleus with orbiting particles, that can attract and repel other particles.

Possibly, the expansion rate of electrons and atoms, is different in these other realities of different interdimensional densities, including also the dream world and the world of thought, which are also made of atoms and molecules according to Seth, although somehow of a different nature.

More likely though, the expansion rate of the electrons and the atoms, is identical in all realities and dimensions, since the expansion rate of the neutrinos must be universal to all realities and planes of existence. If it were not so, we would never encounter beings from other dimensions, we would quickly outgrow and leave them behind, or they would. All That Is, or God, is made of these same neutrinos, electrons, atoms and molecules, as it is most likely the universe itself, which is the overall higher consciousness that all other consciousnesses comprise. Just like it is for the particles, cells, bacteria and organs of our own body, assuming they all have a consciousness.

If everything is made of the same electrons and atoms, expanding at the same constant rate, then gravity, or the changing distance between expanding objects, must also be the same everywhere, no matter the dimensional or density world. But weight can be different from one realm to another, for various other reasons explained in previous chapters, starting with different interdimensional densities of matter, vibrational rates of matter and energy, and time rates.

It might just be a question of interdimensional density, different spacing between atoms within molecules, and between electrons within atoms. Also, different spacing between clusters and clouds of electrons. Making all these particles either smaller or larger, vibrating and orbiting either slower or faster, defining other dimensional or density worlds, which affect weight and buoyancy. All so that in dreams, and in higher dimensions, we can easily fly.

It is the orbiting speed, and the size of the orbits, of the smaller particles linking together the larger ones, that define the density of matter and the dimension in which matter exists. It is not the intrinsic expansion of electrons and atoms, at a rate that is constant everywhere, which most probably never changes.

Although I do wonder. If an atom can become larger or smaller, and switch to another dimension, possibly it might proportionally expand slightly faster or slower than our normal atoms. But any discrepancy would quickly, in time, make the physical worlds in higher dimensions, so much larger than ours, that it is unlikely. It is said that other dimensional worlds are superimposed upon our third dimensional world, they exist right here right now, sharing the same space as us, and they always did.

This conundrum is easily resolved. It is not the size of the object that matters, or the size of the particles. It is the overall expansion rate of matter, which is always constant. It is true that the Earth will proportionally expand much more rapidly than a ball free floating in space above it, but their expansion rate is the same at all times.

The strange creatures sometimes spotted, such as the Mothman and the Bigfoot, are very likely existing in a lower dimensional world sharing our own world. And sometimes they find themselves in our physical reality. Or in unusual circumstances, both our worlds are vibrating at a similar frequency, and then we can see and interact with them. Just like UFOs, ultraterrestrials, extraterrestrials, spirits and ghosts, may be living in a higher density world that sometimes we can see and interact with.

As John A. Keel says in his books, possibly those aliens are not from outer space, they are from Earth, but all their particles and physical reality, vibrate at a higher frequency than ours, all their particles orbit faster, and they live faster. For a long time, as mentioned earlier, they may have been worried we will eventually find a way to see them, interfere and exist in their reality. So they have been lying to us about who they truly are and where they come from.

I highly recommend you read his books *The Mothman Prophecies* and *The Eighth Tower: On Ultraterrestrials and the Superspectrum.* Also *Operation Trojan Horse: The Classic Breakthrough Study of UFOs,* and *Flying Saucer to the Center of Your Mind: Selected Writings of John A. Keel.* I believe he is getting very close to the truth, he certainly confirmed a lot of my theories and ideas.

I like his idea of the electromagnetic superspectrum, that takes into account the difference in frequencies, relatively speaking, from one dimension or density to another. And the reference to ultraterrestrials, the distinction that aliens are not extraterrestrials in nature, that they are from the fourth dimension or higher, also living on Earth in a physical world just like ours. Imagine if we were to suddenly develop the technology to see their cities.

There is a scene in the movie *Valerian and the City of a Thousand Planets,* that shows how this works, when they visit the planet Kirian, a massive extra-dimensional bazaar called "Big Market". They wear coloured glasses to see the higher dimensional world, and they use converters to move in and out of the higher density world. The entire bazaar scene is worth watching, it shows exactly what I mean and the kind of technology we could develop in the future.

According to Seth, this pulsation, or different interdimensional frequency rates between physical realities, is due to how fast matter is being replaced or replenished by our brain, in the construction of our reality. Someone vibrating at a higher frequency, and suddenly capable of communicating with the dead, might very well simply be someone replenishing the matter of their bubble reality much faster than others. Simply because they can channel in, more energy than others from the top of their head, rendering them capable of replacing the matter of our reality more often and faster, raising the frequency of the matter and of themselves in the process.

It still does not tell us how we can achieve this, but in a psychological reality, it seems logical that you simply need to think about it to make it happen. As it is through thoughts, and possibly through visualising and verbalising our thoughts, that everything happens. It is hard to tell, but I will have a better idea once I have finished reading *The Early Sessions* of Jane Roberts, and all her books. It was mentioned that Seth will describe, in the subsequent books in the series, how the molecular structure of other planes of existence differs from our physical reality.

However, I think it is fairly certain that the expansion rate of electrons and atoms remains the same in all realities. Because it is all the same energy that builds up all these worlds, and these worlds are all still made of matter, and apparently, they are all superimposed and intertwined. Consequently, the very same laws of physics must apply to all, and now they can.

In Universal Relativity, whenever we are in motion our speed is always zero. We shrink from other frames of reference we would be moving away from, and we expand compared to other frames of reference we would be moving towards. Indeed, we never really move if our own bubble virtual reality is being projected in 3D right from inside of us.

I have been thinking and visualising this for decades, and now I feel pretty confident that I am correct with Universal Relativity. It is the logical conclusion, of motion being relative to the point of view. And everyone under deep hypnosis, in the books of Dolores Cannon, talk about travel to other solar systems and galaxies, and even time travel, as being instant, and a question of resonance and the frequency of matter.

Even Seth, in Jane Roberts, insists that we must forget building a spaceship that will take decades to reach another planet in another solar system, as this is not the way we will travel in the future. Especially if, somehow, Atomic Expansion Theory is recognised, and we witness another industrial revolution.

I can only think that the frequency of matter, or the right resonance, which is unique to each place in the universe at any particular time, involves the speed of the orbiting neutrinos, and the size of their orbits, within and between electrons. And the same for electrons within and between atoms and molecules. The speed of these neutrinos and electrons, defines the size of these particles and their vibrational rate. And the expansion and contraction of matter is motion, instant motion anywhere in the universe at any particular time in our history, even the past or the future, even if they are constantly changing.

Everyone is creating their own personal bubble reality, independently from each other, following patterns or blueprints our collective consciousness created, like within a network, and continue to change with every new thought, from every contributing mind or consciousness in the universe. Possibly it is easy for aliens, or these ultraterrestrials, to make us see whatever they want, influence our thoughts and create hallucinations. In the end, it is as real as our reality.

And there is not even any other frame of reference that exists, since only ourselves can see and experience our very own reality. When my father sees me in his reality, I, and/or other witnesses, have telepathically sent him the data or information, as to my approximative location, appearance and velocity, so through his mind, he can create me and perceive me in the periphery of his own bubble universe.

Seth's expansion of the dream world, and of the world of our imagination, the universe of our thoughts or concept-ideas, and also the expansion of our reality, and by extension of our mind, since it is mostly a psychological construction, is all about how we create mentally these worlds.

These three worlds of dreams, thoughts and reality, which are all different planes of existence according to Seth, meaning they are three different universes we create and interact with, are all working on the very same principle, from a physics point of view, the principle of expansion.

And this expansion is the expansion of matter, because that psychic/mental energy, that pure energy flowing through the universe, cannot be made of a magical substance that becomes the matter of our reality. It has to be made of neutrinos and electrons, which then go on to form clusters, clouds, atoms, molecules, objects and realities.

And what about anti-gravity? How does it work, if there are no gravitational waves or some gravitational force that we could counteract, since gravity is now explained by the simple inner expansion of matter? Well, changing the density of an object, practically propels it upwards above the denser atmosphere, without much need for any kind of propulsion, even if this means moving to a higher dimension. Therefore, it is not exactly anti-gravity, but the result is the same.

One important fact, out of Atomic Expansion Theory, is that there is no such thing as a speed of light limit. Which means, that if we can find a mean of propulsion, that can continue to accelerate us for quite a long time, using electromagnetism for example, we can travel at several times the speed of light without being crushed. As long as our acceleration and deceleration are small and gradual. Nothing is stopping us right now from reaching all these other solar systems within reasonable timeframes. It might not be the way to travel in the future, but for now it will do. The Universe is our oyster.

TEN

Atomic Expansion Theory Articles by Roland Michel Tremblay

These articles, published between 2010 and 2016 on my website *www.themarginal.com,* and on various independent news websites around the world, are solely about *The Final Theory*. They were reviewed and edited by Mark McCutcheon, and he fully agrees with their content.

10.1 Expansion Theory - Our Best Candidate for a Final Theory of Everything?

Figure 21 *The Final Theory* by Mark McCutcheon, book cover

On 4 March 2010, New Scientist magazine published an article entitled "Knowing the mind of God: Seven theories of everything", where Michael Marshall reviewed the most promising candidates for the Theory of Everything, the Holy Grail of theoretical physics. In the end, there was no solid conclusion as to which, if any, may lead to this final theory. Each is quite different from the others, demonstrating that there is still no fundamental physical or theoretical agreement on the operation of our universe, and all still fall under the general umbrella of our known scientific paradigm, or Standard Theory.

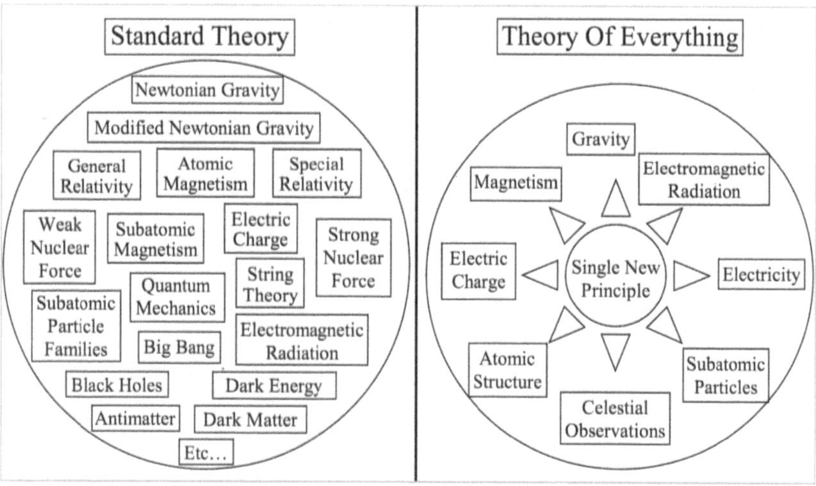

Figure 22 Today's patchwork of theories vs. the Theory of Everything

Yet, this grand final theory is expected to provide a clarifying simplicity and understanding, that is unknown today, implying that it may even lie outside our Standard-Theory umbrella. What if the answer is much simpler and more straightforward, than any of the current proposals, perhaps even lying right underfoot?

This final theory should unite all four fundamental forces (gravity, electromagnetism, and both strong and weak nuclear forces); identify a fundamental principle or particle that does this, and you are well on your way. According to Mark McCutcheon, a Canadian-born electrical engineer and science author, the stable and ubiquitous electron is just such a particle - provided that it operates on a fundamental principle of constant subatomic expansion, rather than today's endless, unchanging "charge".

$$D' = \frac{D - n^2 X_A \cdot (R_1 + R_2)}{1 + n^2 X_A}$$

Figure 23 The Atomic and Subatomic Expansion Equation to calculate gravity: the changing distance (D prime) equals the original distance (D), minus the number of seconds square (n square) multiplied by the Universal Expansion Rate of all matter (X_A or 0.00000077 of proportional size per second square), further multiplied by, in brackets, the radii of the two objects we're measuring the changing distance for (R_1 plus R_2), and overall divided by a denominator calculating the further relative decrease in distance (one plus n square multiplied by X_A).

This switch from "charge" to "expansion", termed Expansion Theory, has surprisingly far–reaching implications, not only for electric charge itself, but also for the nature of the atom and subatomic particles, atomic bonds, magnetism, electromagnetic radiation and gravity. As such, this singular new concept, offers potential scientific explanations, for all known forms of matter and energy, offering further solutions, to the puzzling mysteries and paradoxes inherent in such theories as Quantum Mechanics and Special/General Relativity – the very reason we seek a final Theory of Everything.

This certainly qualifies as thinking outside of known science, as may ultimately be required for a final theory, but is it science? To sincerely answer this question, we must equally apply it to today's theories as well; there must be no free passes on such important issues.

Consider gravity, simultaneously one of the most common yet

mysterious phenomena in our science. Is it a force, as Newton claimed, with no clear reason why it should attract rather than repel, no known power source, and which still puzzles scientists searching for speculative "graviton particles", presumed to mediate its force? Or, despite this most widespread conceptualisation, both taught and used today, even in our space programmes, is it instead Einstein's "warped spacetime" – an entirely different physical explanation, spawning its own puzzles and searches for equally speculative "gravity waves"?

Even the very concept of "dark matter", arose to address a tenfold discrepancy between current gravitational theory and cosmic observations – mysterious invisible matter, that neither emits, absorbs, blocks or reflects any type of radiation, yet is now presumed to be the dominant component and gravitational influence in the universe.

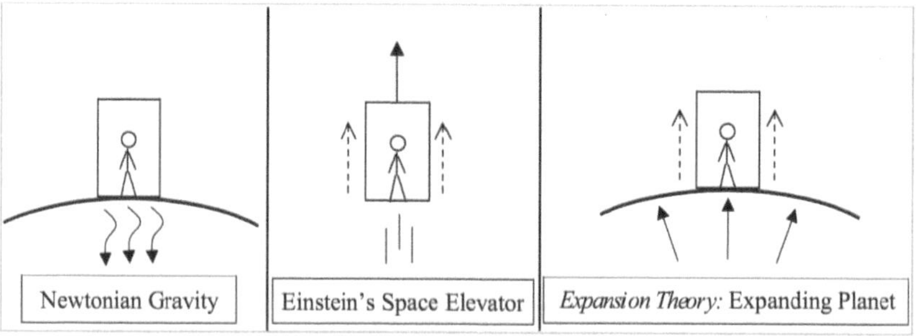

Figure 24 (2-3) Progression of ideas leading to Atomic Expansion Theory

But if we consider the expanding-electron concept, which in turn leads to equally expanding atoms, a new gravitational theory emerges, that actually mirrors Einstein's famous elevator-in-space thought experiment, where standing on Earth, is entirely equivalent to being accelerated upward in space. The force we feel underfoot, is then due to our resulting expanding planet, with dropped objects all equally approached by the ground, rather than the other way around, while the underlying expansion is unseen, as everything expands equally, maintaining constant (relative) sizes.

This would create the appearance of a force, somehow holding us to the ground and pulling all objects equally downward, regardless of mass, just as Newton proposed. And while Einstein opted for "warped spacetime", atomic expansion suggests this far simpler and more literal possibility.

Intriguing perhaps, and while Expansion Theory does provide compelling parallel explanations for many observations, are there any cutting experiments that might set it apart for validation purposes? Consider holding one object, while another of equal mass hangs from it by an elastic band, then letting go. According to Newton, a gravitational force acts equally on all components, accelerating the entire balanced system of two objects and a stretched elastic downward.

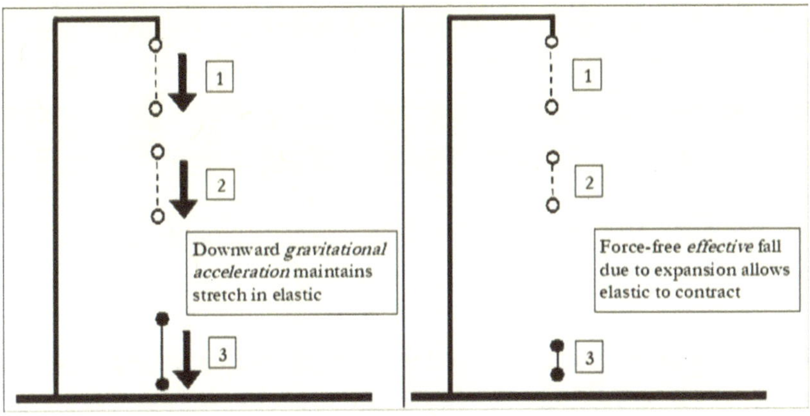

Figure 25 A dropped stretched elastic band should not contract with Newton or Einstein, but should in Atomic Expansion Theory, in a force-free effective fall

Letting go does not free the elastic to contract, but instead frees the entire system to accelerate, with the bottom mass pulled downward, and the resisting inertial mass of the top object now in tow, maintaining the stretch in the elastic, caused by the earlier hanging mass. The gravitational pull, also on the top object, merely matches that on the bottom object, to ensure its mass can also attain the same acceleration, rather than slowing the fall of the overall system, with the stretched elastic then still remaining.

But this is not what happens. The elastic actually contracts during the fall, pulling the objects together. Yet, this should not occur, according to either Newton's gravitational force, or Einstein's "warped spacetime". However, it should occur, if the planet's expansion was initially pushing the held object upward, forcefully stretching the elastic before the drop – an influence that would vanish during free-fall, which allows the elastic to contract, as everything floats free while the ground approaches.

This simple cutting experiment, would appear to seriously challenge both Newton and Einstein, according to the Scientific Method, where even a single negative result disproves any theory, while supporting the expanding-atom concept of gravity. But this would also appear to raise serious questions about Einstein's theories of relativity, since Einstein's "warped spacetime" concept of gravity, hails from his General Relativity theory, which in turn follows on from his earlier Special Relativity theory. Is this really possible?

Consider the famous "Twin Paradox" thought experiment, where a speeding astronaut returns to Earth, to discover he is much younger than his Earthbound twin. A logical flaw in this paradox claim, has been reluctantly but increasingly acknowledged over the years, since "everything is relative" in Special Relativity theory, so either twin could be considered speeding or stationary, removing any absolute age difference. But, should this flaw be pointed out, focus is invariably switched away from Special Relativity, since only the astronaut underwent actual physical acceleration in his travels, which is instead the realm of General Relativity. This switch is generally presented as a resolution to the issue – but is it?

First, this switch to General Relativity, invalidates the still often-claimed support for Special Relativity, from both this famous thought experiment, and from all related physical experiments, such as speeding particles in accelerators, or atomic clocks on circling airplanes or satellites. Yet, this fact is typically neither discussed nor even acknowledged, leaving many with the impression that the Twin Paradox, and related physical experiments, still fully apply to and support Special Relativity theory.

Second, even the switch to General Relativity, appears to be a flawed solution to this issue. One of the cornerstones of General Relativity, is the Principle of Equivalence, which states that the acceleration due to gravity on Earth, is entirely equivalent to being accelerated through space at an equivalent rate – no experiment should be able to discern any difference. This means that, even though this acceleration would produce near-light speeds within months, there should still be no physical difference between this scenario, and that of standing on Earth the whole while.

So, according to both the "everything is relative" aspect of Special Relativity, and the Principle of Equivalence in General Relativity, there would appear to be no such phenomenon as "relativistic time dilation", despite widespread citation, of iconic theoretical and experimental claims to the contrary. Not only would this seem to question some central claims of Special Relativity, but doubly so for General Relativity, considering the earlier drop test as well. And notably, the expanding matter concept, differs not only with the drop-test prediction of both General Relativity and Newtonian gravity, but also with the time dilation claims, related to Special and General Relativity, providing very different explanations of these scenarios.

Interestingly, another test of this new concept of gravity, would be to weigh an object directly on the surface of the far side of the Moon. Since the Moon is about a quarter the size of Earth, its expansion-based surface gravity, would be one quarter as well. This is also calculated by Newton's mass-based gravitational equations, before revising lunar mass assumptions, to match direct surface measurements from our space programmes. And while the actual one-sixth surface gravity – only directly measured on the near side, and presumed to extend around the lunar surface – is currently explained by assuming a less dense lunar composition throughout, there is now another possible explanation.

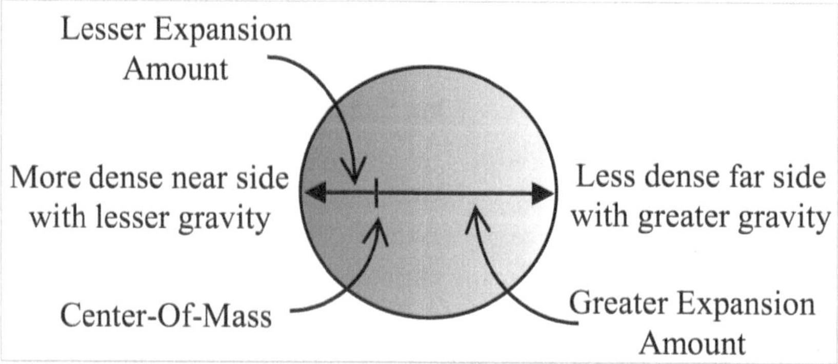

Figure 26 (3-14) The moon's non-uniform density causes differing gravity

Expansion Theory suggests a varying density, from most dense on the near side, to least dense on the far side, which is also in keeping with one of the commonly proposed lunar creation scenarios. In this case, since the expansion of objects, would proceed from their centre of mass, there would be less expansion force on the near side, and more on the far side, due to the resulting off-centre expansion. This suggests double the surface gravity, or one third the gravity of the Earth, on the far side, to average to the one-quarter gravity suggested by the Moon's size. This fact would not affect either the Moon's shape or any orbits about it, but could only be determined by direct surface contact.

Atomic expansion also means, that ocean tides cannot arise from a lunar influence. Tides must occur from internal dynamics within Earth – an inner wobble, that in fact must exist according to classical physics, since the centre of mass, of the overall Earth-Moon rotational system, lies off-centre within our planet. This view suggests why the passing Moon coincides with rising tides, roughly speaking, but for purely internal reasons that follow from the creation, evolution and ongoing dynamics of the Earth-Moon system.

One of the most celebrated successes of Newton's gravitational-force theory, and a milestone in our science, is the extension of Earth's surface gravity, to a forceful "action-at-a-distance" quality. Newton claimed this force reaches out into space, holding the Moon in orbit. But this proposal, not only still has no solid physical explanation, for how it might operate – 300 years later, but also offers no explanation, for the immense and endless power source, that must exist to support such a powerful undiminishing force. We have developed conceptual abstractions to address this issue, in the absence of solid physical explanations, but this has left us with an array of speculative gravitational theories and physical explanations.

In contrast, the expanding atom concept, explains orbits at a distance, as an inescapable geometric consequence of surface gravity. It is easy to see, for example, how dropped objects would effectively fall, due to planetary expansion alone, and how horizontally tossed objects, would similarly curve and plummet toward the ground. Such dramatic momentum change, solely due to the geometry of expansion, demonstrates that gentler curving trajectories, traversing increasing fractions of Earth's circumference, would result with greater horizontal speed. Unlike the absolute straight-line momentum, suggested by Newton's first law, there is actually no reason, such an object would not travel one-third, one-half, and eventually a full orbital circumference, about an expanding planet, as its speed increased.

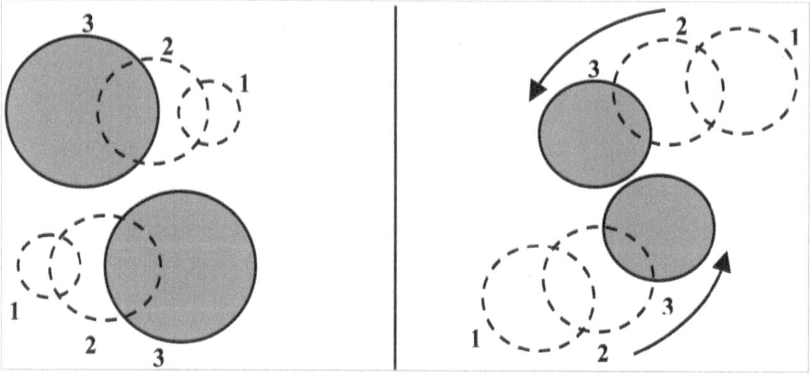

Figure 27 (3-6) Concept (left) and result (right) of the Natural Orbit Effect

Atomic expansion suggests additional explanations, for observations throughout our solar system, such as planetary orbits and interplanetary space travel. Consider two planets passing each other, while their expansion closes the gap between them. We would never actually see such expansion directly, as a size change, if we and all other objects expand equally, maintaining constant (relative) sizes. So the closing gap between the objects, could only manifest as unchanging planets, and for some reason, curving toward each other while passing. Newton suggested the reason is a still-unexplained attracting force, while Einstein instead proposed four-dimensional warped spacetime. However, curves and orbits would also follow, quite naturally and unavoidably, from the pure geometry of expanding matter alone.

The dynamics of orbiting expanding moons and planets, would also result in the entire solar system, and all of its contained orbits, expanding as well. This can be shown to explain, such occurrences as gravity assist manoeuvres, that accelerate spaceships as they pass planets – and where there are no known g-forces in the process – an otherwise mysterious manoeuvre, that lacks proper explanation today, upon closer examination.

And, at the level of the overall solar system, this expansion addresses widely known puzzling anomalies, with the Pioneer space probes and other spacecraft, as they travel through the solar system and beyond. These deviations from predicted trajectories, can now be considered as possible artefacts of our Newtonian gravitational models, based on a force emanating from a given mass, rather than the geometry of expansion.

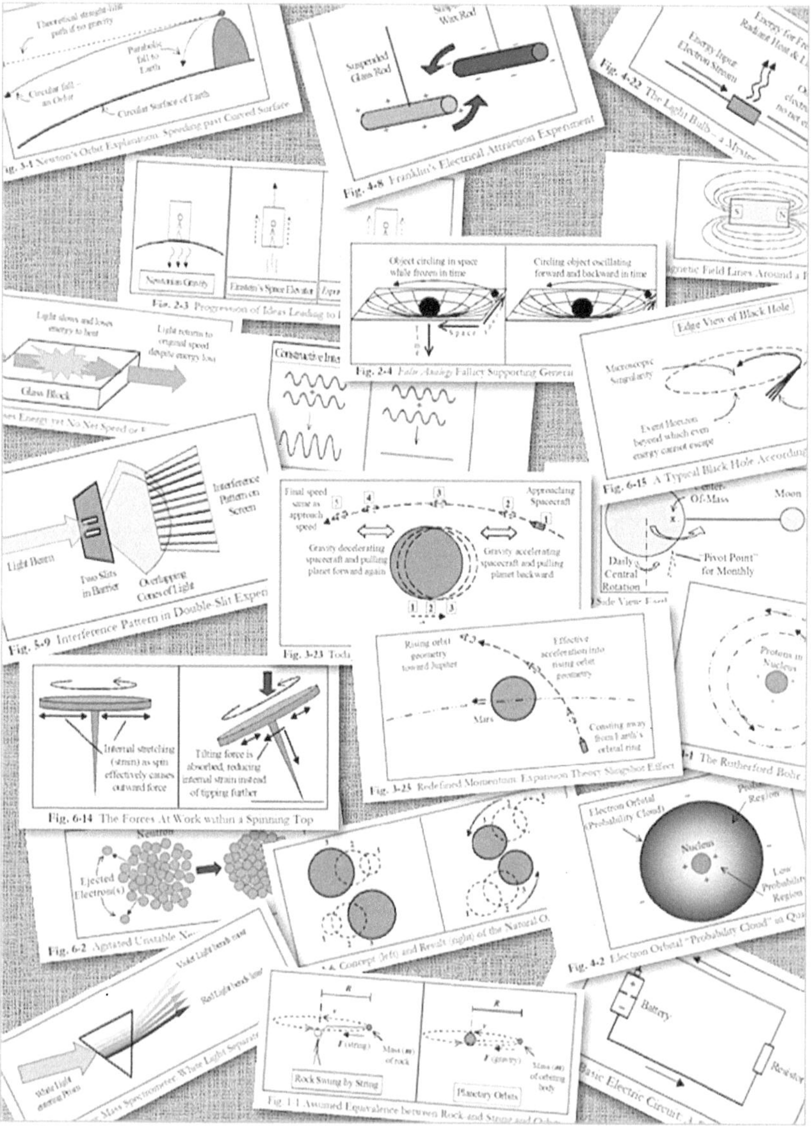

Figure 28 Collage of figures from the book *The Final Theory* by Mark McCutcheon

And, much as expanding atoms replace the notion of "gravitational energy", expanding subatomic particles, replace the energies of "electric charge" and "strong and weak nuclear forces". These separate energy concepts, similarly become unnecessary abstractions, in an atomic model where neutrons and protons are not true particles. Instead, they are clusters of expanding (not "charged") electrons, while "orbiting" electrons bounce repeatedly off, the resultant continually expanding nucleus.

Today's "strong nuclear force", holding the powerfully repelling "positively charged" nuclear protons together (whose required power sources are both oddly absent), is replaced by the crushing force, of rapidly expanding protons and neutrons against each other.

And the "weak nuclear force", causing occasional nuclear decay, further suggests the characterisation of neutrons, as less stable clusters of active expanding electrons, that occasionally eject an electron, to become a more stable proton cluster. This is a more straightforward proposal, for this nuclear "decay" process.

This concept extends further to chemical bonds, currently attributed to endless electric-charge, or electromagnetic energy. And even beyond, as external clouds of expanding electrons, that we call electric and magnetic fields. Even electromagnetic energy, such as heat and light, becomes clusters of freely expanding electrons, pushing one another through space. While electricity is expanding electrons, pushing each other through wires, and extending outward as a surrounding magnetic field.

In the end, all known forms of matter and energy, become manifestations of the singular unifying phenomenon of expanding matter. Although easy dismissals are tempting with most alternate theories, a closer look may well show Expansion Theory, to be much more scientifically viable, comprehensible and verifiable, than the other seven "theory of everything" candidates. In fact, such a comparison could be very eye opening indeed.

Mark McCutcheon is the author of The Final Theory: Rethinking Our Scientific Legacy. *For further reading on Expansion Theory, visit www.thefinaltheory.com. For more information about this new revolution in science, read these excerpts from his book:*

Pioneer Anomaly, Slingshot Effect and Gravitational Inconsistencies Explained
www.themarginal.com/pioneer_anomaly.pdf
The Final Theory by Mark McCutcheon - *Chapter 1 - Investigating Gravity*
www.themarginal.com/FinalTheoryChapter1.pdf

10.2 Breakthrough in Faster-Than-Light Travel and Communication, and the Search for Extraterrestrial Intelligence (SETI)

Interstellar space-travel, and near-instant communication faster-than-light; discovering a network of intelligent extraterrestrial signals; harnessing the mysterious instantaneous quantum-entanglement effect. These are all either science fiction, or things we will probably never live to see or understand, correct? Not at all. By the end of this article, you will see how clarifying, a simple but extremely fundamental misunderstanding in our science legacy, makes all of this a viable reality – now.

10.2a All Just a Misunderstanding

How can this be? It's not as surprising as it may seem, but follows from the leap of understanding that often occurs, when simple misconceptions are clarified. It is just that this particular misconception, reaches back centuries, to the very nature of matter and energy.

Misunderstanding the nature of light, for example, the physics underlying "quantum mechanics", and the meaning of experimental results, can easily produce a strangely complex science, and an oddly bizarre and paradoxical universe. But is this a true reflection of the world around us, or is something else going on here?

In actuality, much of today's science, emerged in much simpler times centuries ago, now forming a legacy, of often unquestioned and presumed truths about our world. But on closer examination, many of these presumed truths, are actually just abstract models, and not physical answers at all. This misconception is powerfully reinforced, in our educational systems and science programmes, locking us into an often-troubled science paradigm of abstractions, contradictions, mysteries and paradoxes.

Newton, for example, only created a mathematical model of his proposed gravitational force. He offered no scientific or physical explanation, for its still-unknown source of power, its law-violating undiminishing pull across eons of time, or how and why it even attracts matter together at all. Einstein offered a radically different, even more abstract and mathematical model for gravity, two centuries later. He provided even fewer practical answers, resulting in both models now residing in our science.

But can the singular physical nature of gravity, truly be captured by two different theories? Can light from a distant source, be simultaneously both a "wave of pure energy", and a "quantum-mechanical photon particle", only physically "choosing" one or the other, based on how it is later observed? Can a magnet cling energetically to a fridge, against the constant pull of gravity, yet need no explanation for this endless energy?

10.2b Clearing it All Up

So, what is the centuries-old misunderstanding in our science? As mentioned earlier, it turns out to be a simple misunderstanding of the nature of matter and energy. Today we think of matter as passive lumps of mass, with various ethereal energy phenomena, actively driving everything. But what if, instead, it is matter itself that is active – both atomic and subatomic matter – and there are no separate "energy" phenomena at all?

The simplest example of this, is a rethink of gravity, where all atoms actively expand very slowly and in unison. Nothing would appear any different over time, but standing on an enormous expanding planet, means we would certainly feel this expansion beneath us – as a force pushing upward under our feet. Also, held objects would feel heavy, as we essentially carried them along with us, while being pushed upward. And they would appear to be pulled to the ground when released, actually allowing the expanding planet to strike them instead.

All objects would have to "fall" at the same rate, regardless of mass, which is precisely what does occur. Tossed objects, would similarly appear pulled to the ground in curving paths, that extend further the faster they are tossed, eventually never reaching the ground at all, but continuing around the planet in a continual orbit, if tossed fast enough.

10.2c Quantum Entanglement Explained, and a Communications Revolution Revealed

Today's science explains quantum entanglement, as an experimental observation, where two photons from the same light source travel together, then are sent on two separate paths, yet apparently maintain a mysterious link with each other. Thus, if one is later altered (such as a change in polarisation), the other is instantaneously altered in the same fashion, no matter how far apart they may be. This is considered a mysterious faster-than-light communication, between two "entangled photons".

However, with the new understanding, the nature of light is radically changed, from separate photons fired through space, to continuous beams of expanding subatomic-matter clusters, that our eyes detect, to generate the experience of colour and brightness. In this case, this is not an experiment with two photons, exhibiting mysterious "quantum entanglement". Instead, it is merely two separate unseen continuous beams, of expanding matter clusters, physically connected back to where they were split, from one initial beam. Then, the more likely explanation of the "entanglement" effect, is that an influence altering one beam, is conducted along this continuous span of unseen physically connected matter clusters, to affect the other.

And, since vibrations in solid objects, travel faster the denser the material, the speed of conduction, through the extremely dense span of such subatomic-matter clusters in light, may well be extremely rapid – even far exceeding the speed of light. The "entanglement" experiments appear to suggest this possibility, of conducting signals along beams of light, at speeds that so far appear to be instantaneous, providing a practical possibility for faster-than-light communication.

Crucially, advanced species would likely use such communication, along existing beams of starlight, rather than generating light or radio waves, and waiting for them to physically cross space, at the relatively slow speed of light.

An analogy for the difference between these two signal-transmission methods, can be seen in the desktop toy, with a line of hanging metal spheres, suspended next to one another, often called Newton's Cradle. When one sphere is pulled back, then released to swing and strike the others, a sphere at the far end is immediately ejected. A long line of such spheres, would allow transmission of such a signal to the far end, far faster than it would take, for a single sphere to swing that same distance on its own.

Likewise, the new understanding suggests, that we might develop ways to look for such rapidly conducted signals, hidden within existing starlight, that already connects us with the distant stars. Rather than today's method, of looking for conventional light-speed signals, as embedded features that move along with the beam.

There could well be a hidden interstellar Newton's Cradle-style internet, all around us, awaiting any civilisation, that reaches this fundamental understanding of matter and energy. We could also find a way, to conduct such a signal within the light of our Sun and Moon, back and forth with the Earth, revolutionising telecommunications in the process.

10.2d Much More to Come

This new understanding rethinks everything, showing that even space travel, is no longer limited by Einstein's claimed "speed of light limit". Such apparent limits, from particle accelerator evidence, simply stem from our misunderstanding of the true nature and behaviour, of the accelerating "magnetic and electric field energy". These "energy" fields, are actually fields of expanding subatomic particles which, by nature, expand at the speed of light. Hence, particles in these accelerators, could never possibly go faster, when powered by such means.

Also, apart from the practical propulsion challenges, the fact that our spacecraft have never come close to light-speed, has nothing to do with such a speed limit in nature. It is more because we haven't truly tried, since we believe today's light-speed myths.

The power-source violations, of the "law of conservation of energy", by gravity, magnetism, and many other observations, also now vanish. "Quantum mechanics" is as a mere fanciful model, for a much simpler physical manifestation, of expanding subatomic matter. And it suggests viable and simple new advances, toward detecting entire networks, of intelligent faster-than-light extraterrestrial communication, conducted along existing starlight.

Gone are the physical mysteries and confusion, of "quantum entanglement" and "quantum paradoxes". It turns out, they never were true physical mysteries at all, but mere human misconceptions, of a far simpler physical reality right under our noses. The need for communication satellites circling the globe, may well now be a thing of the past, and real-time robotic "virtual reality" exploration, of distant moons and planets, could be a reality for us, controlled instantaneously from here on Earth.

Is such a revolution worth studying, considering how many billions we are spending today, on telecommunication research, quantum computers, satellites, SETI and other technology? I think so. And maybe, using light instead of our actual technology, could solve the bandwidth issues, and might produce less radiation, so it would be healthier.

10.3 Dark-Matter, Dark-Energy and the Big-Bang All Finally Resolved

10.3a The Crisis in Cosmology

Today's crisis in Cosmology is perhaps best demonstrated, by an apparently accelerating expansion of the universe, where a 'Dark Energy' must be postulated, to justify this extraordinary acceleration apart – an energy that itself defies both explanation and the Law of Conservation of Energy.

And the crisis only deepens, considering there would have to be between 5 and 50 times more matter in the universe, for Einstein's gravitational calculations to match observations. Which is why unseen 'Dark Matter', was conjectured to keep these calculations "correct", and account for the "missing mass".

A further reason for this crisis, is the now familiar 'Big Bang' theory – the current consensus belief, backed by the attendant vested interests, and therefore largely unquestioned, but which actually fails under objective analysis, showing a universe that is not expanding apart at all.

Objective observation shows a universe, where billions of stars organise into inwardly spiralling galaxies, that group into larger stable Galactic Clusters, then further into enormous Super Clusters, that thread throughout the universe, providing definition even on the grandest scales.

The fact that one camp solidly and consistently, reports this stable observational structure of our universe on all scales, while a separate camp powerfully and enthusiastically, promotes a completely incompatible "Big Bang" / "Dark Energy" ever-accelerating universe, merely reinforces the enormity of the crisis in today's Cosmological community.

10.3b Deepening the Crisis: Painting the Wrong Picture of Our Universe

However, despite the enormity of this crisis, it can be readily resolved once we identify where it all began – a fundamental flaw in Hubble's Law, which incorrectly assumes, that redshifts observed in starlight shifted toward lower frequencies, correspond to velocity away through space. But first, it is worth taking a brief overview of the journey that brought things to this point.

Earth was once considered flat and at the centre of the universe, until it was found to be round, and in a Sun-centred solar system, as only a small part of a huge galaxy. And even our galaxy, the Milky Way, was later found to be one out of billions of galaxies, in our immense universe.

Meanwhile, the universe itself changed from three dimensions to presumably four – once time was included, and from entirely regular matter, to apparently mostly invisible matter filling the cosmos. It even changed from a static universe to one coasting apart, and now even a shocking accelerating expansion.

This creates a picture of a universe, composed of a literal 'four-dimensional spacetime fabric', bursting forth from an actual 'Big Bang' creation event. With unseen exotic physical 'Dark Matter' filling the universe, and a new form of unexplained energy – a mysterious 'Dark Energy', repelling everything apart ever-faster.

To counterbalance increasing acknowledgement, of the complete lack of solid physical and scientific grounding, for much of this picture, is a unified front of increasingly fortified scientific consensus, and continually growing Nobel Prize support.

This process has resulted in a number of key assumptions and theories, becoming effective 'laws of nature', after which, by definition, observations must fall in line, and not conflict to suggest other interpretations. And while this is an important process for scientific advancement, it can potentially entrench incorrect 'laws of physics' into our science, for indefinite periods of time, sometimes with disastrous results.

Indeed, even suggesting conflicting interpretations, once a 'law' was established, was a very dangerous act, that history shows often carried severe penalties. It is important to note that today's science, has its own tight control and dismissal mechanisms, that can indefinitely entrench detrimental 'laws', for reasons of vested interest, just as effectively as in times gone by.

10.3c Resolving the Crisis: Where It All Began - "Hubble's Law"

One such example is Edwin Hubble's assumption, nearly a century ago, that an observed redshift in starlight to lower frequencies, indicates a star's motion away from us in space – based on a simple analogy, to the known Doppler Shift of moving sound sources in air. This Doppler-like assumption, was made at a time when light was presumed to be a wavelike phenomenon similar to sound, and when there was far more interest, in the enormous cosmological implications of Hubble's assumption, than the actual immense differences between light and sound.

Sound, for example, is simple compression waves, conducted at the speed of sound, in an air medium. Whereas, even in Hubble's day, light was considered a somewhat mysterious 'electromagnetic energy wave', that somehow always travelled at constant light-speed – and with no conducting medium at all. Further, light was increasingly considered an even more mysterious quantum-mechanical phenomenon, that is somehow simultaneously also a 'photon particle', only settling on either wave or particle once detected.

Despite these serious problems, with Hubble's initial Doppler-inspired 'redshift equals velocity' assumption, the intrigue and controversy, created by a possible expanding universe coasting from a 'Big Bang' creation event, tipped the scales, entrenching both "Hubble's Law" and this radically new cosmological picture, into our science. The increasing observations of redshifted starlight all around us, now had to align with Hubble's apparent 'law of nature'. Which could now only mean, that everything was moving away and apart, locking cosmology into this line of thought ever since.

So powerful was this view, that it now dominates our understanding of the universe, despite the fact that light, is nothing like Doppler Shift-able sound waves. Also, that light is easily red-shifted, merely by passing it through materials such as common plastics. Given this fact, the redshifts observed in starlight, across millions of light-years of space, filled with all manner of materials and gases, might not be particularly surprising. Redshifts could simply indicate a great distance across space, and not a Doppler-like velocity at all.

10.3d The Problems with Hubble's Law Deepen

One of the most critical problems, with Hubble's "redshift equals velocity" claim, is that it contains a clearly fatal logical and physical error, that has been overlooked for nearly a century now. If the universe were actually expanding as Hubble claimed, it would produce nothing like the straight-line, regular spacing, of the associated Hubble-Law diagram. As the plot progresses to ever-greater distances, it also represents observations that are ever further back in time as well.

The universe is now believed to be about 14-billion years old, with billions of galaxies dotted throughout it, at such great distances, that we can only reasonably describe them in terms of light years – the distance light travels in a full year. Even the nearest galaxies are millions of light-years from us, with most of them billions of light-years away, across the observable universe, extending 14 billion light-years in all directions.

As such, the points plotted on the diagram below, represent

redshift measurements and the associated velocities, as required by "Hubble's Law", for galaxies at observed distances of one billion light-years, two billion light-years, three billion light-years, etc. And, of course, these are presumed velocities, that were occurring one billion years ago, two billion years ago, three billion years ago, etc., since it took that long for their light to reach us.

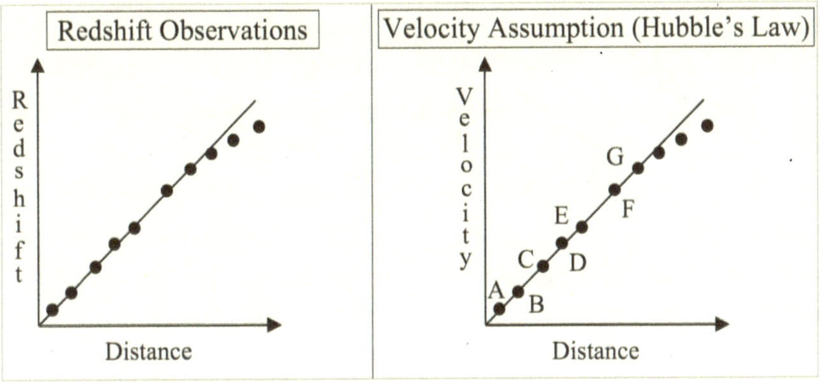

Figure 29 (6-13) Hubble's Law – The "redshift=velocity" assumption

Cosmologists are well aware of this, frequently stating that looking out into space, is equivalent to looking back in time. Yet, they have failed to follow this understanding to its inevitable, troubling conclusion. Galaxy A, spotted one billion light years away, and which was apparently traveling at its redshift-indicated speed, one billion years ago, will in actuality now be far more distant, as it continued speeding away, over the intervening billion years.

Any regularly spaced plot of galaxies, along Hubble's straight line, where both redshifts and velocities increase linearly with distance, shows galaxy spacing that existed in the past, and which must now be spaced with ever-increasing gaps out from us – in the present state of the universe.

This effect would be even more accentuated by, for example, the third galaxy out, Galaxy C, spotted three billion light-years away. Its "Hubble redshift" speed, is supposedly three times faster than

Galaxy A. Thus, by the time of this observation, it would have been speeding for three times longer than Galaxy A, making its present gaps with the other galaxies greater, by a far more disproportionate amount than shown.

So, although the diagram shows gaps that all appear fairly equal in size, and expand apart fairly equally as well, to give a uniform universe from any location – as required by the so-called Cosmological Principle – this is actually not at all the case. Hubble's "redshift equals velocity" interpretation, actually describes an impossible universe, where the gaps grow disproportionately larger with distance – from the perspective of every galaxy in the universe.

But of course, it is logically and physically impossible, for the actual present-moment gaps, to be ever-larger outward from us toward distant galaxies, while also being simultaneously ever-larger outward, from distant galaxies toward us.

This impossible, but very real paradox in today's Cosmology, is shown below, with two completely incompatible gaps from galaxies A to G, viewed from Earth, and back again from galaxies G to A, viewed from galaxy G. The only possible conclusion, is that Hubble's Law is incorrect.

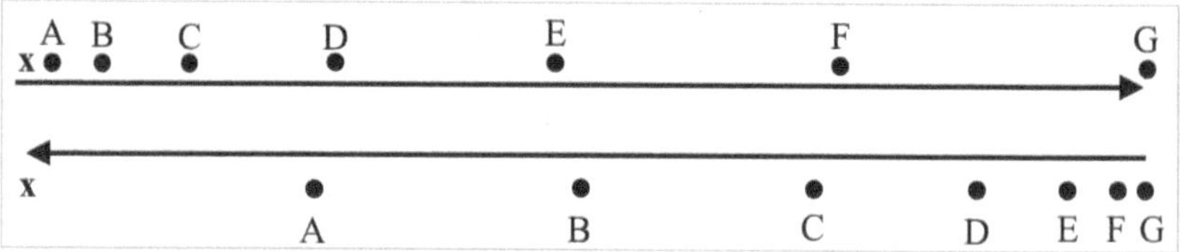

Figure 30 (6-14) Galaxy spacing out from us and from a distant galaxy

10.3e Erroneous "Dark Energy" Invention Draws Nobel Prize

It is this type of problem, that has been building within the Cosmological community, to crisis proportions. Critically so in recent years, over the issue of specific types of supernovas and their distances, and apparent speeds of recession away from us. Chronic fundamental oversights in Hubble's Law redshift interpretations, and logical paradoxes in misinterpretations, have led cosmologists to conclude, that supernova evidence proves our universe is accelerating ever-faster, due to a mysterious form of "Dark Energy", that is entirely new to science.

Despite the fact that this new form of energy has no scientific explanation, has never been demonstrated in any experiment, and has never been identified on any energy spectrum, its "discovery" roughly 14 billion light-years away, via the spectra of a handful of supernovas, was recently awarded a Nobel Prize.

However, none of these paradoxes or mysteries would exist at all, if the universe were relatively static, and the detected supernova brightness and spectral redshifts, merely arose from the nature and distance of the enormous spans of intervening space, rather than "Hubble's Law".

10.3f Further Crisis Resolution: Einstein's Erroneous General Relativity Theory

Einstein's General Relativity theory presents a similar issue, with Einstein's reputation helping elevate it, also, to an effective gravitational 'law of nature'. He modelled the universe as a 'warped four-dimensional spacetime realm', rather than one of gravitational forces, in regular three-dimensional space.

This effective 'law', has likewise required observational interpretations to align with it, for nearly a century. It led to cosmologists, inventing physically unexplained and completely undetectable 'Dark Matter', that neither emits, absorbs, reflects or blocks light, to account for tenfold discrepancies, between Einstein's theory and observations.

However, were it not for this 'law of nature' status, and Einstein's reputation, following proper Scientific Method, would merely have led to the conclusion, that Einstein's largely untested theory is simply wrong. It is now verified to be out by an enormous factor of ten, when simply held to the same objective unbiased scientific observation, and scrutiny, as any other theory.

A bitter pill to swallow, for huge vested interests in the scientific community, who have staunchly supported this picture of the universe for decades? No doubt. An embarrassment, to see a scientific icon knocked from his pedestal, with one of his most long-standing revered theories, shown to be completely false? Definitely. Reasons to knowingly send the whole of science and humanity off-track indefinitely, to keep these facts hidden? Hopefully not!

Now, much as with "Hubble's Law", once we allow ourselves to question Einstein's effective 'law of nature', and simply hold it up to the same scientific scrutiny as any other theory, its tenfold disagreement with observations, immediately disproves it.

And, just as letting go of this communal mental block, frees us to completely eliminate the mysterious 'Dark Energy', attached to our "Hubble's Law" beliefs, it also frees us to eliminate the mysterious 'Dark Matter', invisibly dominating our universe, attached to our General Relativity beliefs.

10.3g The Ongoing "Cosmological-Constant Blunder"

Einstein created his General Relativity theory – a merger of Newton's gravitational-force theory, and Minkowski's four-dimensional spacetime abstraction – to try to provide a truly universal model, and new physical understanding of gravity, due to his strong dissatisfaction with Newton's theory; hence today's 'warped spacetime' notion of gravity was born.

However, finding that his resulting equations, could not be used to describe the static universe, generally presumed at the time, but only one that either expanded apart or contracted together, Einstein further merged a sizeably altered version of his equations, describing a hypothetical mass-less universe, envisioned by Willem de Sitter.

Since de Sitter had already added an arbitrary control parameter, to Einstein's equations, in order to tune the dynamics of his hypothetical universe, Einstein adopted this parameter, later called the 'Cosmological Constant'. He was hoping he might set it to a value, that made his equations valid, for our presumably static universe.

But during his attempts to model a static universe, with these merged equations, Einstein became convinced, that the universe was actually coasting apart, based on Hubble's 'redshift equals velocity' interpretation, of an observed redshift in starlight, all around us.

Famously calling his arbitrary 'Cosmological Constant' introduction, his "greatest blunder", Einstein removed it from his General Relativity equations, in the hope that his original equations, might better model a universe, now apparently coasting apart, presumably from a 'Big Bang' creation event.

But cosmologists later noted, that observations based on Hubble's 'redshift equals velocity' assumption, actually suggest that the universe is not only coasting apart, but actively accelerating apart ever faster, apparently driven by a mysterious repulsive 'Dark Energy', now dominating the universe.

And, since even Einstein's return to his original equations, could not model this accelerating expansion apart, his "Cosmological-Constant blunder" removal, is now being reconsidered for return to General Relativity theory.

This time, its arbitrary addition is intended to model an accelerating universe model, that hopefully works for this current belief. It is now persuasively renamed from Einstein's "greatest blunder" – his so-called 'Cosmological Constant' – to the apparently new and mysterious 'Dark Energy', pervading the universe.

10.3h General Relativity - a Theory that has Never Actually Worked

The problems from all of these arbitrary abstractions, mergers, additions, removals and re-additions, have steadily mounted. Newton's 'gravitational force' theory, has actually never been scientifically explained, despite its familiar and intuitive nature. And neither has Minkowski's 'spacetime' abstraction, which Einstein merged with it, to create his General Relativity theory.

Further, de Sitter never claimed that his hypothetical mass-less universe, with its arbitrary 'Cosmological Constant', was to be taken literally. And nor did Hubble ever scientifically explain or validate, his 'redshift equals velocity' assumption, that compelled Einstein to later remove his 'Cosmological Constant blunder'.

As a result, and considering further ongoing alterations of Einstein's General Relativity equations, by various scientific camps, we have had a core theory of gravity for nearly a century now, that has been cobbled together, and repeatedly and arbitrarily altered, to try to match the latest observations and beliefs, yet which has never actually worked at any point – a fact that remains the case even today.

This is the very reason, for the seemingly endless stream of 'mysteries' and 'surprises' and 'puzzles', that seem to arise from Cosmology, decade after decade. In actuality, it is not our universe that is so strange and bizarre, but merely the distorted theories and beliefs, through which we view our universe, that make it appear so.

10.3i False Supporting Evidence: The Cosmic Microwave Background Radiation

Even further cracks appear, once we begin allowing ourselves to question today's cosmological picture. For example, it can be readily shown, that faint 'Cosmic Microwave Background Radiation', arriving from space, is not the 'Big Bang whisper', that it was claimed to be decades ago. It is merely microwave noise, from our local solar system and galaxy.

It is now known, that the early ground-based detector was far too crude, to discern any faint patterns from outside our galaxy. And that, the featureless detected radiation, contained no inherent indication of a distance of origin. Yet it was considered, and still is, the first detection and mapping, of the structure of the early universe. This remains the case despite hindsight now showing, that this early 'Big Bang whisper' claim, is an obvious error that clearly should be retracted.

Once again, the case cannot be overstated. The original crude ground-based detector, initially stumbled into an unexpected random microwave hiss of noise. It was eventually decided by some, that this hiss was the highly sought-after proof, of the then controversial 'Big Bang' theory. After which patterns, presumably representing the structure of the early universe, were said to be found within this radiation. A Nobel Prize was even later awarded to this effort.

Crucially, it was much more quietly later acknowledged, that the original random microwave hiss, could only have been just that – a meaningless random hiss. This is because, an extremely advanced detector would have been required, to discern any meaningful pattern, from a severely diminished signal, across billions of light years of space, then across the 100-thousand light-years of our active galaxy, then through the radiation of our solar system and its burning sun, and finally our dense atmosphere.

Also because, the detector in question was orders-of-magnitude too crude - only able to pick up a meaningless random hiss of microwave noise, given the task just described, to detect any signal originating outside our galaxy, or even well within it, let alone from the distant early universe.

Nevertheless, even today, despite full realisation and recognition of the above, no retraction of the original erroneous "Big Bang whisper detection", has ever been issued from the Cosmological community. In fact, in quite the opposite move, a more detailed 'Cosmic Microwave Background Radiation' detection, performed from orbit, is said to agree with the initial 'early universe' detection pattern. This is despite recognition, that the initial pattern is now verifiably meaningless – with the new detection effort also awarded a Nobel Prize.

10.3j Erroneous Double Nobel Prize-Winning 'Big-Bang' Proof

Today's now largely unquestioned 'Big Bang' theory, was originally heavily debated, until ground-based radio telescopes detected background microwave hiss. It was claimed to have patterns, identifying it as ancient, greatly redshifted radiation from the 'Big Bang' creation event – drawing a Nobel Prize.

However, it is first important to note, that this background microwave hiss, is quite unlike redshifted starlight. It is not associated, with any observable distant stellar objects, whose radiation, is dramatically redshifted down to microwave frequencies. It is instead, an almost perfectly uniform hiss of background microwaves, arriving from all directions. As such, it is no more evidence of an origin of ancient radiation, from the distant early universe, than recent microwave noise, generated from the billions of stars in our local galaxy, or even our nearby blazing Sun.

Indeed, we now know, from the far more sensitive COBE satellite, that the original detected radiation, was composed almost entirely, of radiation from precisely these local sources. COBE also showed that any patterns, that may exist in the faint radiation from beyond our galaxy, would be far below the detection threshold, of the original radio telescopes. And so, the initial Nobel Prize-winning claims, of patterns from the early universe, were verifiably, nothing more than wishful thinking at best.

Secondly, the COBE, and later WMAP satellites, also showed that, any true ancient radiation patterns, would be dwarfed by combined microwave disturbances and noise, a hundred-thousand times more powerful, in crossing the immensity of intergalactic space, then our own galaxy of billions of active stellar objects, then our solar system with its blazing sun, and finally our highly absorbing and distorting atmosphere.

And, since a great deal of this overwhelming distortion, is largely or completely random, there is no way to reliably characterise and extract it, to uncover any extremely subtle, and highly distorted inter-mingled patterns, a hundred-thousand times weaker.

Despite these facts, those behind the COBE and WMAP satellite projects, claim that not only have they clearly discerned even more detailed patterns, of the early universe, from this radiation, but also that these patterns correlate with those in the original detection claim, drawing yet a second Nobel Prize.

Yet, as just described, it is a physical impossibility, to recover and reconstruct any faint original signal, from the overwhelming distorting random noise. Also, these later projects actually show that, it would have been impossible, for the original detector, to discern any actual 'early universe' patterns whatsoever, in the original radiation.

So, the first scientifically responsible outcome from the COBE and WMAP projects, should have been a resounding retraction, of the initial Nobel Prize-winning claim. As the technology to make such a claim, was now unquestionably lacking – by orders of magnitude.

However, not only was no such retraction made, but instead, the verifiably meaningless "structural map of the early universe", was re-released, after the COBE and WMAP data had been processed for months, until it was convincingly superimposed on top of it, reinforcing it with further detail, and collecting a second Nobel Prize in the process.

As a result, a meaningless noise signal is, even today, held as verification of the 'Big Bang' theory, cementing it into our science, our collective psyches and belief systems, to the point where it is now a largely forgone conclusion, and unquestioned – if not even unquestionable – scientific 'fact'.

10.3k Time to End Our Mounting Theoretical and Physical Crisis in Cosmology

So, from a theoretical perspective, our core gravitational theory in Cosmology, General Relativity, is a patchwork of scientifically unexplained, abstract sub-theories, with a 'Cosmological Constant' that is continually added and removed, in repeatedly failed attempts to match observations, and proclaimed as everything from a “great blunder”, to a mysterious ‘Dark Energy’ permeating the universe.

And from a physical perspective, we have recent claims, of a universe somehow accelerating apart, after a presumed 'Big Bang' creation event, despite conflicting observations increasingly showing, all the stars existing within stable galaxies or galactic clusters, threading throughout the universe.

The recent law-violating claims, of a universe accelerating apart, are based on Hubble's largely unquestioned, and scientifically unverified assumption, that redshifted starlight equals velocity. And the best 'Big Bang' evidence, is now actually verifiably, erroneously Nobel Prize-awarded microwave noise.

This is undeniably the current state of Cosmology today – and the current destination of billions of public tax dollars, earmarked for scientific investigation and advancement. It is clear that vested interests, in the scientific community, are not about to enact any significant change to this state of affairs. So, it is up to an informed and concerned public, to do something about this ongoing state of crisis, in our science.

10.3l Farewell 'Big Bang', 'Dark Matter', 'Dark Energy' and 'Spacetime'

If we simply allow ourselves, to take a critical look, at a double Nobel Prize-winning observational claim, and re-think two highly questionable century-old ‘laws of nature’, we remove three of today's largest mysteries from Cosmology: the 'Big Bang', 'Dark Matter' and 'Dark Energy'.

It is worth noting, that these Cosmological claims, 'laws' and observations, are largely abstract or remote in nature. And so, they are far more susceptible to being thrown wildly off track, and require extra care and scientific due-diligence.

However, now with appropriate corrective analysis, there is no longer a mysterious infinitely small singularity, from which the entire universe burst forth, no longer completely undetectable exotic 'Dark Matter', dominating our universe, and no longer a mysterious law-violating 'Dark Energy', accelerating the universe apart.

In their place, is a possibly static universe, of potentially infinite size and age, within which, stars of regular matter, undergo continual births and deaths, with gravity-driven dynamics, in ordinary three-dimensional space.

This leaves a number of immediate questions: Does the scientific community, for some reason, want to retain our current cosmological picture, with its deep and possibly irresolvable ongoing mysteries, and unquestioned "laws of nature"? And if not, and they are truly side-tracked on a centuries-old journey, in search for answers, then, what might this gravity be, that is driving our simple and possibly static and endless universe?

Newton's gravitational-force theory has many problems, as Einstein recognised in trying to replace it. Einstein's warped spacetime theory, has even greater issues. And we certainly won't get anywhere, inventing "Dark Matter" or "Dark Energy". So, what is the answer? The answer is, we need a credible new Theory of Everything, including a new theory of gravity.

For more information about these major issues in our science, read the excerpt Cosmology in Crisis *from* The Final Theory *by Mark McCutcheon:*

www.themarginal.com/CosmologyInCrisis.pdf

10.4 Gravity Breakthrough: Springing into a Gravitational Revolution (written by Mark McCutcheon, edited by Roland Michel Tremblay)

Gravity is one of the most familiar everyday phenomena, yet it has mystified scientists and laymen for centuries. Even today, although the current official position on gravity, is a continual "spacetime warping" around objects – a claim from Einstein's General Relativity theory, it is also still widely considered an endless attracting force emanating from objects, as claimed in Newton's gravitational theory.

Setting aside for the moment the troubling implications, of two different physical descriptions of gravity in our science, it turns out that the behaviour of a simple spring, may hold the final answer to this age-old mystery.

Consider what happens, when a loosely coiled spring is stretched apart from both ends, while laying on a tabletop, as shown below in the left-hand frame. The opposing forces spread equally across the spring, causing an equal coil spacing across the spring, which also occurs, whether either force pulls fully from the very end, or is divided to pull directly on each coil:

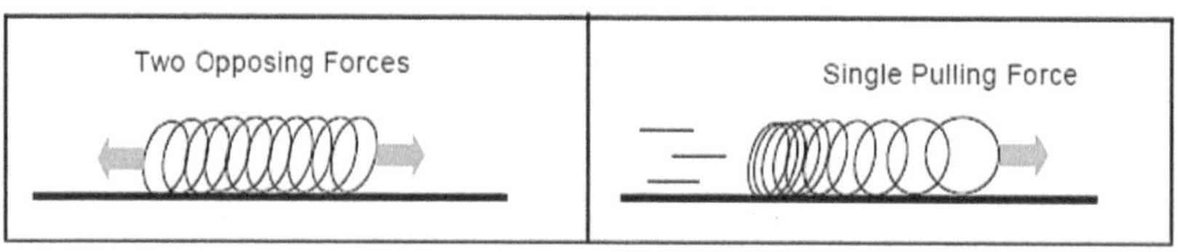

Figure 31 Two opposing forces, stretching a spring horizontally, compared to a single pulling force

However, with only a single continual pulling force on one end, shown on the right, the coils stretch more at the leading end, as they strain to continually accelerate, the ongoing resisting inertia of the rest of the spring. In this case, there is successively less stretch toward the trailing end, as there is successively less trailing-coil mass, to cause inertial drag.

This deceptively simple experiment, has enormous implications, for both Newton's gravitational force, and Einstein's 'warped spacetime' theory of gravity – and for understanding the true physical nature of gravity itself.

The first important point, is that it highlights a widely overlooked but critical error, surrounding Einstein's famous "space elevator" thought experiment, which forms the foundation of his Principle of Equivalence, and his later associated General Relativity theory of gravity.

10.4a The Erroneous "Principle of Equivalence"

Einstein claimed that all experiences and experiments, occurring inside a constantly accelerating elevator, moving upward in deep space – far from any gravitational influence – would be indistinguishable, from them occurring under the influence of Newtonian gravity on Earth.

This claim is known as the Principle of Equivalence, and forms the cornerstone of gravitational physics in today's science. However, the simple spring experiments just discussed, can be used to show that this is an erroneous claim, with enormous implications for our understanding of gravity.

Similar to the left-frame tabletop experiment above, a hanging spring on Earth, should have two opposing forces distributed across it, equally spreading its coils – the force of gravity pulling downward, and the restraining force, that effectively pulls upward.

However, as in the right-frame of the above tabletop experiment, a spring attached to the ceiling, of Einstein's continually accelerating deep-space elevator, far from Earthly gravity, should exhibit the unequal coil distribution of a spring, pulled from only one end:

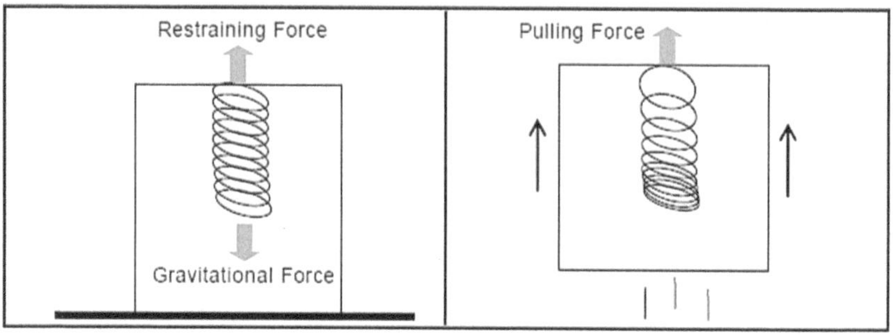

Figure 32 A hanging spring disproves Einstein's Principle of Equivalence with Newton

So, this shows that Einstein's claimed "Principle of Equivalence", between Newtonian gravity, and pure acceleration in deep space, must be wrong – the effect of being accelerated upward in space, must differ from an attracting force emanating from a planet.

If Einstein had remained faithful to his original "space elevator" inspiration, rather than developing his General Relativity theory for equivalence to Newton, he would have produced a new understanding of gravitational physics, that clearly differed from Newton's, and which could be easily tested by a simple hanging spring experiment.

Instead, Einstein effectively abandoned his space-elevator inspiration, in favour of a mistaken "Principle of Equivalence" to Newton, and a related "warped space-time" proposal for the physics of gravity, in his General Relativity theory.

10.4b A Verifiable Revolution in our Understanding of Gravity

But why concern ourselves with this hanging spring issue, in a deep-space elevator, especially if we already know that Einstein's Principle of Equivalence, and General Relativity theory, are widely accepted today, and supposedly even proven, by highly sophisticated experiments?

The reason is, this very same hanging spring experiment can be performed by anyone – by simply suspending a well-known spring toy from one end, showing that gravity on Earth behaves precisely, as in Einstein's original space-elevator inspiration, and not as in either Newton's "gravitational force" theory, or Einstein's equivalent "warped space-time" General Relativity theory.

This simple experiment, shows a hanging spring with an unequal distribution – here on Earth – which could only occur, if it were continually accelerated upward from its suspended end, and not stretched uniformly by an attracting "gravitational force", or equivalent "space-time warping".

This further shows why no solid scientific explanation, for the operation of Newton's proposed attracting force, has ever been settled upon, and nor has its apparently endless power source ever been identified or explained.

This also means, that Einstein's efforts to mirror Newtonian gravitational theory, in his General Relativity theory, are equally verifiably in error. And that the experiments presented as proof, were conceived and designed, such that their claimed "success" actually constitutes no particularly meaningful result at all.

10.4c Could the Evidence Still Support Today's Gravitational Theories?

The preceding discussion, shows that Newton's theory of an attracting gravitational force, is readily disproven by a simple hanging spring. And so is Einstein's 'warped space-time' General Relativity theory, which was deliberately designed, to be functionally equivalent. But before addressing what all of this means, it can still be tempting to dismiss the above discussion, with intuitive support for today's gravitational theories, such as the following:

'The coils at the top of a hanging spring, simply bear the weight of the rest of the spring hanging below. And those further down, have fewer coils below them, thus less weight to bear, stretching successively less. This results in more stretch at the top, and successively less toward the bottom – a non-uniform hanging spring.'

This may initially sound reasonable enough, but the first hint of a flaw in this logic, is that it is at odds with the earlier tabletop experiment, showing that two opposing forces (such as gravity pulling down, and a restraining force pulling up), should result in uniform coil spacing.

So, what is the logical flaw in the above reasoning? It is the presumption that the strain caused by weight, is solely due to a downward pull from gravity, and that this strain accumulates with the weight of the lower coils, tallying to greatly stretch the upper ones.

The error in this logic, is shown in the first frame of the diagram below, where an object's weight is shown as solely due to a downward pull from gravity. If it were literally true, that there is nothing but a downward force on the object, then the object would not rest as a weight in our hand, but would be in a weightless free-fall, as shown in the second frame.

The very reason the object is not in a weightless free-fall, but sits instead as a weight in our hand, is because there is an opposing force – in this case from our muscles – holding it in place, as shown in the last frame:

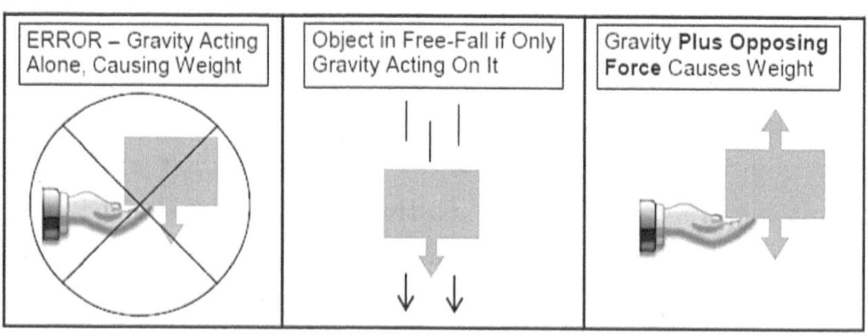

Figure 33 Gravity plus opposing force causes weight

Similarly, the error of both logic and physics, in the weight-based reasoning for the non-uniform hanging spring, is the suggestion that the weight of each coil, is solely due to a gravitational force (frame 1 below), with downward weight accumulating along the spring.

In actuality, a scenario with only a downward gravitational force, would produce a spring in weightless free-fall (frame 2 below), which would accelerate toward the ground, with no stretching at all, in the absence of an opposing upward force.

A statically hanging spring (last frame), however, actually has two opposing forces distributed throughout it – according to today's gravitational theory (gravity acting downward, and the restraining force acting upward), which, again, should equally spread its coils.

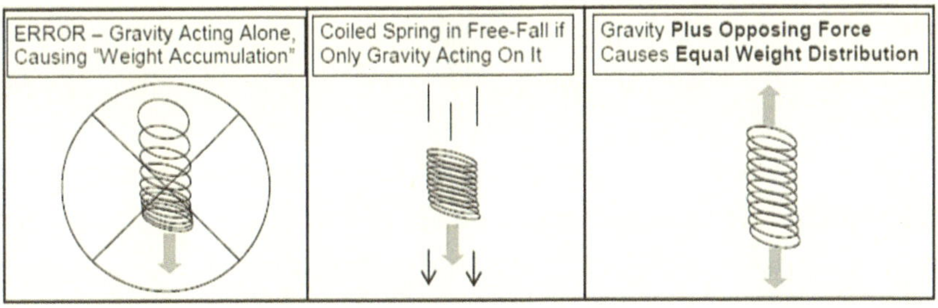

Figure 34 Gravity plus opposing force, should cause equal weight distribution in the coils of a spring

There can be no such thing, as 'accumulating coil weights' in a hanging spring, caused by a lone gravitational force pulling them downward, and adding up to cause a non-uniform distribution. There can only be equally stretched coils, from two opposing forces.

There remains no viable explanation in today's science, for the observed non-uniform distribution of a simple hanging spring – experimentally disproving all current gravitational theory.

10.4d The True Nature of Gravity Finally Revealed

So then, what does all of this mean? If a simple hanging spring, experimentally disproves both Newton's attracting-force suggestion, and Einstein's warped space-time proposal, what does it mean, when the experiment mirrors Einstein's upwardly accelerating space elevator? A strong hint, is that this experimental result, is completely in line with a compelling new theory of gravity, known as Expansion Theory.

This new theory states that all atoms – and, by extension, all objects composed of atoms – are slowly and continually expanding, by roughly one-millionth their size each second. This underlying expansion is unseen directly, as everything expands equally, but is felt as a force beneath us, from our huge expanding planet. But it is seen indirectly, as all objects, regardless of mass, appear to fall equally to the ground (which actually rises to meet them all equally).

This explains why Einstein's space elevator, correctly captures the observed behaviour of a hanging spring on Earth, since our planet's constant expansion, effectively acts as an elevator constantly accelerating us upward.

In this case, a suspended spring on Earth, effectively hangs in the "elevator", with a singular continual upward pulling force, as we hold it suspended. Here, the accumulated ongoing resisting inertia, of the lower coils, would indeed cause greater stretching in the upper coils, and result in the non-uniform distribution observed in the hanging spring.

This also explains, why the spring's behaviour does not match, either Newton's or Einstein's demonstrably flawed downward-pulling theories of gravity, which could only cause equal coil distribution.

And, according to the formal Scientific Method, any single solid contrary experimental result, definitively disproves any theory – regardless of how well it may otherwise match or model observations.

The only viable conclusion from this discussion, and from both experiment and our understanding of physics, is that the effect we call 'gravity', arises from a universe of actively expanding matter, rather than one of separate inert matter, and active "gravitational energy", with no known, and necessarily draining, power source.

Ultimately, in Expansion Theory, all forms of "energy" turn out to be, various forms of actively expanding atomic or subatomic matter, with "energy" being a mere misunderstanding, of a universe where all matter actively expands, by its very nature of existence.

10.5 Revolutionary new physics could lead to ultimate weapons of mass destruction

As a sci-fi author and science consultant, for films and television documentaries, I have made it my mission to seek out every alternate theory out there – no matter how crazy they may seem – to explore any possibility of new physics, which might be uncovered and put to use. That is, until I came across something so perfect and convincing, that I had to stop in my tracks.

I now realise, to my complete astonishment, that the true Theory of Everything *already exists*, in a book published soon after the new millennium – you just haven't heard about it yet. Who is this author, Mark McCutcheon, and what is this book, *The Final Theory?* My God! This is not the usual crackpot theory, used as fodder for some lame sci-fi TV series; this is it – the first truly viable new physics to have ever arisen.

As I read on and on, for the first time gaining a complete understanding, of all that is currently mysterious and weird in theoretical physics, including Newton's gravity, Einstein's relativity, and especially the quantum mechanics of Niels Bohr, I found myself making a complete turnaround. I will never see the world the same way again.

Yes, this Theory of Everything really *does* explain it all – no stone is left unturned. It explains everything, up to the mysteries surrounding the Pioneer satellites leaving the solar system, and the difficulties we have encountered landing spaceships, on other moons and planets. At long last, complete explanation of everything there ever was in physics, has *finally* arrived – and, as I now realise, the *first* and *only* true understanding we have ever had.

McCutcheon has just rewritten the whole of physics, and it makes sense. I cannot see how he could be wrong, and believe me, I have tried to prove him wrong in a long correspondence with him, now spanning several years.

I then started wondering, how this new understanding of every single phenomenon in physics, might be used to perfect weapons of mass destruction, even build the most powerful nuclear weapons ever made. Up until now, we never truly understood what was behind the physics, that we have exploited to make such weapons possible – it was largely a progression of abstract models, and trial-and-error. Now I believe, through this new theory, that we can have a complete understanding of it all. I cannot fault it; perhaps you can?

Who would have thought one single little idea, could revolutionise the whole of physics, and completely rewrite Newton, Einstein and Bohr, all in one time swoop? That is now the almost mythical hope for the long-sought Theory of Everything, but who among us actually believed it was possible? Completely rewriting the books on all these theories of gravity, relativity and quantum mechanics, all based on one singular new principle, running throughout it all?

Yet, I assure you, this is it; this is what Mark McCutcheon has achieved. I'm sure without even realising, the enormous impact this will have on the world, and how we will go about building more powerful weapons of mass destruction, as a consequence.

Are we wise enough, to handle this new knowledge, this entirely new physics? Well, above all else, we must do our best to uncover and communicate, the whole truth to everyone, to every single student studying physics, mathematics and chemistry world-wide. There is no point going any further, on our current path in science – our Standard Theory, filled with warped spacetime, quantum mysteries and relativity paradoxes.

These are chronic issues in our science, that never disappear, and which no one truly understands – despite the odd academic assurance to the contrary. This is not because the vast majority of us, lacks the intellectual capacity, while the handful, who are heavily invested in these specialties, are, oddly, so far beyond the rest. It is because these concepts, are inherently nonsensical, to any sensible mind without an agenda.

Only now can I see this with crystal clarity. I can now re-read the explanation attempts, offered for the various paradoxes in our science, and see the logical flaws in all of them – the Emperor is finally disrobed. Mark McCutcheon has just rewritten the whole thing for us, and I challenge you to read this book and tell me otherwise.

Now we can have a true understanding, of how nuclear weapons truly work, and what $E=mc^2$ really means. Now we can know the true nature, of subatomic and atomic bonds, furthering research in such areas as biological, chemical and nuclear weapons. I don't know, but with such an understanding, of how physics and chemistry truly work, I think we may have finally stumbled upon the Holy Grail of all wars, but also, the Holy Grail of revolutionary new technologies.

And yet, I cannot say this should be hidden from view. I cannot say that such an understanding of all physics, should be kept secret. I believe we should seek the truth above all else. Especially considering how many billions of dollars, we spend on experiments and studies, that can now be clearly demonstrated to be primitive, misguided or useless, by theory alone.

This strikes me all the more, when I encounter documentaries about the physics of our world, where presumably great theoretical physicists, are actually just wasting time and money. Their time and resources could be far better spent, if they had the real physics to work with, rather than the overbaked convoluted theories they have now, and that they struggle to dumb down for us.

Expansion Theory, as Mark McCutcheon has aptly named this new theory, will ensure we will never struggle to understand how to land a spacecraft, on any moon or planet in the solar system, or needlessly lose one more astronaut. Gravity is not what we think it is today – it finally has a proper, entirely new definition. This is our only viable way out of the solar system, once we decide to seriously get to work on it.

I am now going to give you, for the first time ever, a core insight from the new theory, which you can take or leave as you see fit. But if you read the book, just as I did, I have no doubt you will be convinced of its truth.

And it is simply this: all can be explained by the fact, that the electron is the only fundamental particle in the universe, and that it constantly expands, causing all atoms, which are composed of electrons, to also expand. And since atoms are expanding entities, all objects made of atoms expand as well, at a constant rate confirmed mathematically. Every atom or object in the universe, doubles in size every 19 minutes. But this growth is unseen directly, since everything maintains the same *relative* size.

This means, the Earth expands by 4.9 meters each second, pushing us upward. Falling objects never truly fall, they float in the air until the Earth reaches them whilst expanding. This expansion goes on to explain all of gravity, all orbits, and all of energy. Isn't that revolutionary?

This is Newtonian gravity completely destroyed. Objects are not attracted to each other, by a mysterious gravitational force acting at a distance. The distance between objects simply diminishes, due to the fact that all objects undergo a constant underlying expansion, including of all orbits, while empty space does not.

And as we cannot see this expansion, since we also expand, this results in effectively constant-sized objects, moving toward each other. At which point, they may hit or start orbiting each other, determined by the geometry of expansion behind the scenes.

The expansion of electrons and atoms explains everything: gravity, chemical bonds, a new model of both the electron and the atom, without any inherent charge or magnetism. The entire universe is alive with expanding electrons, pushing against each other, both within and outside the atom, in electron clouds or electron clusters, that explain radiating heat and light. The theory goes on to explain radio waves, and the whole spectrum of energy waves, which are no longer waves at all, but various configurations of expanding electrons.

The whole of physics is now explained by expanding electrons. We now have light without mysterious photons, without Einstein, without quantum mechanics. No more weird claims, that no one can understand or explain.

I never thought in a million years, that one person could come up with such a radical change of the whole of physics, all in one go, all within one revolutionary book. You thought Einstein was a genius, just wait until Mark McCutcheon explodes unto the world of physics world-wide. Every single book of physics, math and chemistry, will have to be re-written.

Can you explain what you observe in the world around you? Planetary orbits, atomic bonds, the configuration and dynamics of all matter in this world? The true nature of all these energy forces, which is shown to be an obsolete concept, in this new theory? Now I can, because I now have a clear understanding of all of physics.

I'm so pleased I won't have to struggle to wrap my brain around understanding Einstein and quantum mechanics. I now understand why I could never quite do it, no matter how hard I tried – because it was all truly nonsense, and thankfully it is all gone now. My God, even Newton's gravitational force is gone, replaced by a much more viable physical explanation.

It is all now simplified completely; there are no more mysteries. Now I can worry instead, about how such an understanding of the true physics of this world, how the true *final* Theory of Everything, could potentially be used to annihilate the world we live in.

We should always remain one step ahead, shouldn't we? I know there are many crackpots out there, coming up with new definitions of just about everything in physics, including gravity. But remember, as in the well-known parable, while many may cry "*wolf!*", there eventually really *was* a wolf. McCutcheon is the one voice in the crowd, who has finally truly struck gold; he cannot possibly be wrong in my opinion, and his book proves it in theory, with so many proofs page after page.

In fact, I am so impressed, I think this is such an important book which will revolutionise everything, that I'm thinking about turning the book into a television documentary. This is how much I believe, it is the only true physics we need to consider. If you read the book and you are interested in financing such a documentary, please contact me. It will happen at any rate, I can assure you. It's just a question of whether you are in, on this coming revolution, or not. This entirely new paradigm truly needs to get around – it *has* to be recognised, so we can stop wasting time, money and energy, on the wrong physics.

I can't believe how hard it is, without a proper marketing machine, to reach out and tell the world about a critical new development. In an ideal world, a book such as this, should have had such an impact by now. But obviously our world is far from ideal, so more must be said and done. You, too, will have no doubt about this, once you read the book.

The Final Theory itself is the only proof we need. It is not possible to read it, and state that this is not it, that "*Expansion Theory*" does not explain everything in physics, unless somehow you feel threatened by such a revolution. The long-awaited revolution in science is here, we now have a true Theory of Everything, for the first time ever. And now let's see the leaps and bounds science can make, when finally, we have a full understanding of what we are doing.

It may lead to better weapons of mass destruction, but we already have efficient ones that could destroy us many times over, and we're still here. More importantly, it could also lead to a new technological revolution, that has been too long in coming. Actually, it is our only hope to instantly solve the energy crisis, which is responsible for most wars in this world, and the only mean we might have, to replace our fleet of petrol and diesel cars and trucks, to electric ones. A new physics for a new millennium, and now we can finally reach out to the stars!

10.6 The Final Theory of Everything - An in-depth interview with the author Mark McCutcheon, By Roland Michel Tremblay

In 2002, at Universal Publishers, Mark McCutcheon published the first edition of his book *The Final Theory*. It was an instant success, however, it also created some controversy. It presented to the world, a new Theory of Everything that worked out of the box, to replace Newton's Law of Universal Gravitation, Einstein's Theories of Relativity, and Quantum Mechanics.

The theory is based on the one principle, that matter simply expands at a constant rate, reducing the distance between objects in an acceleration, explaining gravity. We stay on the ground, because Earth is expanding underneath our feet. This expansion is unseen by us, since everything expands proportionally at the same rate, including our measuring instruments and ourselves.

All particles, like protons and neutrons, are now composed solely of electrons, expanding against each other within the atom, preventing them from flying apart, and elucidating the strong nuclear force. A neutron is an unstable particle by nature, that decays into a more stable proton, releasing an electron in the process, to justify the weak interaction.

There are no more charges or charged particles in electromagnetism, just like there is no more force acting at a distance in gravity. And yet, the entire energy spectrum can be explained, through the simple expansion and movement of the electrons.

They move from where there is a surplus of electrons, toward where there is a depletion, in a search for balance. In a crossover effect, they also grow and shrink as they get excited, equalising in size while pushing against each other, on wires for electricity, in clouds for magnetic fields, in clusters for heat and light, and in compressed bands for electromagnetic fields, such as radio signals and microwaves.

The electrons are bouncing off, between the expanding nuclei of atoms, justifying chemical bonds. Orbits are described, by an ingenious new natural orbit effect, through the simple geometry of expansion.

This is all four main forces of nature explained, while they are no longer forces per se, now that we understand their true nature. We could never have found a Unified Field Theory within Standard Theory, as there was nothing to unify, in these four arbitrarily identified forces of nature.

Mark McCutcheon is Canadian, like me, and has a combined Electrical Engineer / Physics degree. He has successfully re-written our entire physics, and yet it has gone largely ignored, after the initial interest. This is the first in-depth interview with him, to catch up with what he thinks of his Theory of Everything now, and what he feels about the state of our physics today.

Q1. Could you describe to us, in your own words, Atomic Expansion Theory? How is it different than any other individual out there, claiming to have found a Theory of Everything, and how does it differ from the other Theories of Everything, being pursued by theoretical physicists today?

Expansion Theory, which encompasses both expanding subatomic and atomic matter, arose from my increasing awareness of many holes in accepted Standard Theory, as I progressed through my scientific education, continuing even through university. Many of the problems were so fundamental, that I saw no clear resolution, either personally or in the understanding of other scientists, who went through this same system. In retrospect, it required an entirely new perspective, on the fundamental physical sciences, to truly address these key issues, and hence an entirely new Theory of Everything.

Of course, any viable theory, must be validated by real-world observations and experiments, which was my guide and entire purpose in pursuing this issue – to not only equally explain everything in current science, but further explain that which current science did not. I was not driven by, nor interested, in a personal 'pet theory', but truly comprehensive and objectively verifiable answers, founded in solid logic and scientific principles.

Q2. What was your eureka moment? When did you think for the first time, that if matter was simply expanding, an entire new physics could emerge? How did you feel at the time, and what were your circumstances then?

It was a few years after university, following one of the many TV documentaries I had seen, about Einstein's warped 4-dimensional space-time theory of gravity. It struck me just how fanciful and problematic this theory had grown, from Einstein's original intention to revise Newton's theory of gravity, which he clearly felt was in need of a major rethink.

So rather than accepting, being confined within the framework of one or the other, or both of these problematic theories of gravity, I pondered how I would explain my experience of gravity, simply and clearly, to beings who somehow had never experienced it.

First and foremost, I experienced an effective push from the planet below, which I had to continually combat to remain standing upright. The fact that Newton modelled this experience as an attracting force, was secondary at this point, and, again, a problematic one at that (where was the power source? specifically why did it attract – why not repel? how could it never drain, as it orchestrated the dynamics of our world and universe?).

From this push experience, it then followed that, the only way everyone around the globe, could be experiencing the same push from below, is if the entire planet was pushing outward in all directions – i.e., expanding.

The next logical conclusion, was that myself and every other object around me, had to be similarly expanding, since everything remained the same relative size. And this would make sense, from the fact that everything was made of the same atoms, which would then have to be equally expanding atoms.

The more I followed this line of thought, the more it moved from a simple abstract thought experiment, to a viable explanation, of actual real-world experience and observation, continuing through dropped objects, orbits, etc. And it also then occurred to me, that this was quite similar to Einstein's actual simple thought experiment, that evolved over time into its 'warped space-time' form.

That early experiment was his suggestion, that gravity was indistinguishable in every way, to being in a box in space where there was no gravity, while being accelerated from below through space, at a constant 1g acceleration. All experience and experiments, would be indistinguishable from being in that box on Earth.

However, Einstein apparently never considered, that Earth was *literally* that mechanical acceleration from below, as an expanding planet, proceeding instead down the path that led to warped 4-dimensional space-time.

Q3. After that great idea, you set yourself to write a very elaborate and ambitious book, re-writing the entire physics. Could you tell us about how you went about it, how long it took, how much research you did, the entire process?

It turns out I had accumulated, quite a body of unanswered questions and problematic answers throughout my life, from grade school through university, which I eventually just came to accept and move on. So, once I had the expanding atomic-matter concept, issue after issue flooded back to mind, and as I applied my new perspective, I began to find very sensible, satisfactory answers for the first time.

This process continued to the point, where I had to start documenting it, and over a period of several months, I had what might be considered a very rough start, on an annotated first draft, of what continued to grow into *The Final Theory*, published several years later.

Q4. The book was finally published and it was a success, however it created some controversy. You were confronted by people who could not accept, such a radically different way to describe our world, and who could not conceive that neither Newton nor Einstein could be incorrect, although Newton had already been superseded by Einstein. How did you welcome and deal with the success, and the backlash from firm believers in Standard Theory?

I was very pleased, with the enthusiastic response from many individual readers, both in the form of online reviews, and personal email feedback. It was very rewarding to discover, that there were a great many others having similar struggles, to find sensible answers to many fundamental questions, and that this new perspective, gave them viable answers for the first time, as it did for me.

I was also equally surprised by small groups of people, who banded together via chat forums, to mount coordinated efforts, to attack and defame a book that they clearly had not even read, running on pure misconceptions and false assumptions. I found it particularly saddening, that they were trying to mislead as many people as possible, who could have greatly benefitted from the information in the book – material that I believed, and still do, to be their birth-right of understanding, of the world around them and the universe in which they live. Today such efforts are well known, as a fairly standard Internet Troll response, to just about anything that appears on the radar, but at the time I was very surprised and saddened at this dynamic.

Q5. Now that time has passed, are you disillusioned about the state of our science? This hot pursuit for dark matter, dark energy, gravitational waves, graviton particles, the Big Bang, black holes, wormholes, singularities, parallel universes, etc., all of which cannot exist within Atomic Expansion Theory? Where have we gone wrong?

I have come to see that our current scientific community, is going to continue along as usual, deeply steeped in the various concepts you raise in your question. There is too much history, tradition, and both personal and professional investment in that direction, to expect anything to change.

Every so often there is also mounting pressure, to announce success in one or another of these areas after much time, effort and expense. Be it claims of finding a key fundamental particle, of experimental success further supporting an exotic theory, of apparent detection of a new supporting physical phenomena, etc. I'll leave it up to each individual to decide, whether they wish to seriously scrutinize and question these claims, but I believe the information in the book, will greatly assist any such inquiry.

Q6. In most articles these days, concerning dark matter and dark energy, I noticed that they fail to state why it is thought, that such concept-ideas should exist. When considering Atomic Expansion Theory, such concepts are superfluous. Could you tell us why, exactly, we are searching for dark matter and dark energy, and why you believe we won't find them?

Once Expansion Theory is considered, replacing both Newton's and Einstein's closely related models, of mass/energy gravitational theories and mathematics, you can consider the expansion dynamics of matter, rather than a force or energy effect, emanating from mass.

Orbits can then be understood, as geometry of expanding objects instead – whether it's objects orbiting planets, planets orbiting stars, or solar systems swirling within galaxies. This frees us, in today's theories, from the tenfold discrepancies, in mass vs. observed orbital motion. This conflict has scientists, filling in these huge gaps, with claims of mysterious "dark matter", that does not reflect, absorb or emit, any known form of energy or radiation.

In other words, 'dark matter' is completely made up, with no evidence or physical explanation, rather than taking a second look, at current unquestioned theories. This is the exact opposite of the proper Scientific Method – if experiment and observation, clearly disagree with a theory, the theory must be in error. We aren't supposed to instead invent, inexplicable justifications to retain the theory.

The same applies to the idea of 'dark energy', which violates one of the current fundamental laws of physics – the Law of Conservation of Energy. The interpretation of observations, is claimed to show the universe flying apart, with ever-greater acceleration. But without any explanation, for the source of this supposed ever-increasing energy. Plus, of course, every known attracting or repelling force, weakens with distance. But this mysterious 'dark energy', is claimed to strengthen, even as matter presumably accelerates, further and further apart.

All of this arises from an interpretation, of distant shifted light frequencies, as indicating accelerating velocities, when far simpler explanations exist. Such as the known fact, that similar shifts should be expected, simply due to light passing through space filled with gas and plasmas, for billions of years. Such light shifting occurs, even just passing light through gas or mere millimeters of plastic, in lab environments.

There is a concept of Occam's razor, which says the simplest answer is probably the correct one. Why invent physically unexplainable, law-violating 'dark energy', when there are far simpler and more viable answers?

Q7. Would you say that science is closed to new theories and ideas, and that it is difficult, to get any new idea which lays outside Standard Theory, to be considered? What would you change in science today, that could prevent this serious impediment, to identifying a new Theory of Everything worth considering?

I think the issue is largely one of human psychology. We would all like to think of science as an idealistic machine, of pure objective logic and reason – basically the embodiment of the Scientific Method. But the reality, of course, is that this idealistic machine is operated by humans, who are susceptible to personal agendas, cognitive biases, vested interests, logical-fallacies, funding influences and pressures, competition, preconceived ideas, emotion-based decisions, etc.

Given this reality, although there is much well-intentioned rhetoric around, searching for better understanding, and possibly even, a true Theory of Everything that rewrites the textbooks, for the most part, this seems true only to the extent, that it doesn't upset the human element too much.

When it comes down to it, most expert individuals and institutions, aren't truly willing to potentially risk their authority, reputation, status and funding – to consider that, much of what they have studied and professed, may be in question. And the assumption is, that the sought-after understanding, will be in line with all of these human concerns; but what if it is not? That appears to be the real roadblock.

Q8. What do you think of people like me, and others out there, who have picked up your theories, and are developing them further? Are you encouraging everyone to do the same?

Absolutely. This is not my pet theory. If correct, as it appears to be, it is humanity's birth-right of understanding. It belongs to each and every one of us, allowing us to truly understand our world and universe, and make the most of it in whatever ways possible.

Q9. Any last thought you would like to share with us?

Just to take note of the growing global realization, that the 'Internet Trolls' do not have the interests of others at heart. They have their own personal agendas, and are neither sincere nor knowledgeable in their efforts, doing a great disservice to others who allow them undue influence.

A revolutionary new understanding awaits!

About the Author

Roland Michel Tremblay was born in Québec City in 1972, and is a published author of novels, philosophy, essays, poetry and journals, both in France and in the United Kingdom. He has a College Diploma in Human Sciences from the Cégep of Jonquière in Québec, a Bachelor of Arts degree, Specialisation Lettres Françaises, from the University of Ottawa in Ontario, and a Master of Arts degree in Romance Languages and Literatures (French) from the University of London, Birkbeck College, UK. As a career he produced international conferences, and worked as a Clerk in a Crown Court in London. He now lives and writes in an old cottage deep in the heart of the Welsh countryside.

amazon.com/author/roland_michel_tremblay
youtube.com/@Roland_Michel_Tremblay
patreon.com/Roland_Michel_Tremblay
bookbub.com/authors/roland-michel-tremblay
facebook.com/roland.m.tremblay
twitter.com/RolandMTremblay
linkedin.com/in/roland-michel-tremblay
instagram.com/Roland_Michel_Tremblay
pinterest.com/Roland_Michel_Tremblay
snapchat.com/add/rolandmtremblay
tiktok.com/@Roland_Michel_Tremblay

Also by Roland Michel Tremblay

Destructivism: The Path to Self-Destruction

Who are we? Where are we heading? What is the universe and our purpose within it? *Destructivism* is a philosophical discussion about humanity and the world in general, assessing the grand questions of existence and of our way of life.

The first section concerning politics, covers such topics as democracy, capitalism, identity, freedom, war, globalisation, and justice. The ethics section talks about love, hate, life, trust, and fairness. The epistemology section addresses consciousness, dreams, death, soul, and knowledge. While metaphysics discusses virtuality, determinism, time, and a theory of everything unveiling an entirely new physics, to explain the world we live in.

Between 2008 and 2011, several of these essays gathered a lot of attention, as they were published as articles on independent news websites. Despite *Destructivism* not being overtly controversial, it led me to being placed under surveillance with not one, but two vans sitting at all times outside my flat in London.

My landline was making strange clicking noises, my computer was under constant attack and rendered unusable, and I even received veiled threats. But this pales in comparison to my close friend, a writer who helped with the research and inspiration for these articles, being found dead soon after saying she had been poisoned.

Either writing philosophy equates to being a domestic terrorist to be censored, intimidated or eliminated, or there must be something in this book which was frightening some powerful people. Perhaps it was the essays about dystopia, dictatorship, despotism, economics, New World Order, or rigged elections?

It certainly drives the point home, that unless as a society we soon regain our common sense, we will continue to head towards self-destruction. In a way this book was quite prophetic, since less than a decade later, we pretty much achieved self-destruction. *Destructivism* helps uncover how we got there in the first place.

Anna Maria: A Paranormal Science Fiction Mystery

Living in a thatched roof cottage in Richmond Park in West London, clairvoyant Anna Maria investigates unexplained phenomena and mysteries with her neighbour the Duke of Connaught. As their relationship evolves, they experience parallel universes, time travel, precognition, déjà vu, ghosts, and space exploration.

In ***God lives in Chiswick,*** Anna Maria and the Duke of Connaught meet Ms

Barnsworth, a mentally ill patient who claims she's from another universe in which she is sent to a space station around Saturn. She is also apparently capable of creating human beings out of the vacuum of space.

In ***The Marginals,*** our heroes meet two scientists who are studying statistical probabilities, predictions of the future and people's behaviours. When they meet, they realise that they are a perfect match, but don't dare admit to it in order to avoid being the victim of this predestination.

In ***Psi-Star*** we find out about how Anna Maria at 18 learned to cope with being the most psychic woman alive whilst developing quantum computers in Victoria.

In ***Déjà Vu*** they get to fear a Professor from Oxford capable of changing reality at will.

In ***Dead Girl's Song,*** we meet a special ghost from York in quest for the key to the city.

In ***Righteous Citizens of Sidmouth,*** our characters have to fight against a small group of people who denounce online inhabitants who do not conform to their beliefs.

In ***London's Library from the Future,*** it is not easy to distinguish the reality from what books from the future are describing. Should they destroy the career of the Prime Minister or not?

In ***The Uncertainty of King George Varney,*** the existence of some of our descendants is due to a temporal paradox, and the survival of humanity after two world wars depends on Anna Maria finishing the work she started years ago.

In ***The Box on the Seven Dials,*** the Duke of Connaught experiences for himself what mind powers can do to his own personal existence.

In ***The Richmond Park Experiment,*** we find out about other scale universes.

Finally, in ***Bibliotheca Alexandrina in El Cerro Del Pueblo,*** they investigate interdimensional worlds.

Bibliography

[1] D. Cannon, *Conversations With Nostradamus: His Prophecies Explained* (Trilogy), Huntsville, AR: Ozark Mountain Publishing, Inc., 1991-1994.

[2] D. Cannon, *The Convoluted Universe* (5 book series), Huntsville, AR: Ozark Mountain Publishing, Inc., 2001-2015.

[3] D. Cannon, *The Custodians: "Beyond Abduction"*, Huntsville, AR: Ozark Mountain Publishing, Inc., 1998.

[4] R. M. Cluff, *World Top Secret: Our Earth IS Hollow!: The Scientific, Scriptural and Historical Evidence that Our Earth Is Hollow!*, Sun City, AZ: Independently published, 2020.

[5] E. Hicks and J. Hicks, *Ask and It Is Given: Learning to Manifest Your Desires*, Carlsbad, CA: Hay House Inc., 2004.

[6] J. A. Keel, *Flying Saucer to the Center of Your Mind: Selected Writings of John A. Keel*, A. Colvin, Ed., Seattle, WA: Metadisc Books, 2013.

[7] J. A. Keel, *Operation Trojan Horse: The Classic Breakthrough Study of UFOs*, Charlottesville, VA: Anomalist Books, 2013.

[8] J. A. Keel, *THE EIGHTH TOWER: On Ultraterrestrials and the Superspectrum*, Charlottesville, VA: Anomalist Books, 2013.

[9] J. A. Keel, *The Mothman Prophecies*, London: Hodder & Stoughton, 2002.

[10] J. B. Leith, *Genesis of the Space Race: The Inner Earth and the Extra Terrestrials*, Ooltewah, TN: Timestream Pictures & Books, 2015.

[11] M. McCutcheon, *The Final Theory, Rethinking Our Scientific Legacy (Second Edition)*, Irvine, CA: Universal Publishers, Inc., 2010.

[12] W. Penre, *The Wes Penre Papers - A Journey through the multiverse, Levels of Learning*, Eugene, OR: Self-published online, 2011.

[13] J. Roberts, *The Early Sessions of the Seth Material* (9 book series), R. Butts and R. Stack, Eds., Port Washington, NY: New Awareness Network, 1997-2002.

[14] M. Talbot, *The Holographic Universe: The Revolutionary Theory of Reality*, New York, NY: HarperCollins, 2011.

[15] Urantia Foundation, *The Urantia Book: Revealing the Mysteries of God, the Universe, World History, Jesus, and Ourselves*, Chicago, IL: Urantia Foundation, 1955.

www.ingramcontent.com/pod-product-compliance
Lightning Source LLC
LaVergne TN
LVHW091523200926
848483LV00002B/285

* 9 7 8 1 9 1 5 6 3 3 0 2 6 *